The Natural Way of Farming

The Theory and Practice of Green Philosophy

By Masanobu Fukuoka

Translated by Frederic P. Metreaud

Japan Publications, Inc.

Published by JAPAN PUBLICATIONS, INC., Tokyo and New York

Distributors:
UNITED STATES: *Kodansha International/USA, Ltd., through Harper & Row,
Publishers, Inc., 10 East 53rd Street, New York, New York 10022.* SOUTH AMERICA:
Harper & Row, Publishers, Inc., International Department. CANADA: *Fitzhenry &
Whiteside Ltd., 195 Allstate Parkway, Markham, Ontario, L3R 4T8.* MEXICO AND
CENTRAL AMERICA: *HARLA S. A. de C. V., Apartado 30–546, Mexico 4, D. F.*
BRITISH ISLES: *International Book Distributors Ltd., 66 Wood Lane End, Hemel
Hempstead, Herts HP2 4RG.* EUROPEAN CONTINENT: *Fleetbooks, S. A., c/o Feffer
and Simons (Nederland) B. V., 61 Strijkviertel, 3454 PK de Meern, The Netherlands.*
AUSTRALIA AND NEW ZEALAND: *Bookwise International, 1 Jeanes Street, Beverley,
South Australia 5007.* THE FAR EAST AND JAPAN: *Japan Publications Trading Co.,
Ltd., 1–2–1, Sarugaku-cho, Chiyoda-ku, Tokyo 101.*

First edition: October 1985
Second printing: July 1986

LCCC No. 84–81353
ISBN 0–87040–613–2

Printed in Japan

Contents

Preface, 11
Introduction, 15

Anyone Can Be a Quarter-Acre Farmer, 15
"Do-Nothing" Farming, 16
Follow the Workings of Nature, 17
The Illusions of Modern Scientific Farming, 20

1. Ailing Agriculture in an Ailing Age, 25

1. Man Cannot Know Nature, 27

Leave Nature Alone, 27
The "Do-Nothing" Movement, 29

2. The Breakdown of Japanese Agriculture, 30

Life in the Farming Villages of the Past, 30
Disappearance of the Village Philosophy, 31
High Growth and the Farming Population after World War II, 31
How an Impoverished National Agricultural Policy Arose, 33
What Lies Ahead for Modern Agriculture, 35
Is There a Future for Natural Farming?, 35
Science Continues on an Unending Rampage, 36
The Illusions of Science and the Farmer, 37

3. Disappearance of a Natural Diet, 38

Decline in the Quality of Food, 38
Production Costs Are Not Coming Down, 39
Increased Production Has Not Brought Increased Yields, 40
Energy-Wasteful Modern Agriculture, 41
Laying to Waste the Land and Sea, 44

2. The Illusions of Natural Science, 47

1. The Errors of the Human Intellect, 49

Nature Must Not Be Dissected, 49
The Maze of Relative Subjectivity, 52
Non-Discriminating Knowledge, 54

6

2. The Fallacies of Scientific Understanding, 55

    The Limits to Analytical Knowledge, 55
    There Is No Cause-and-Effect in Nature, 57

3. A Critique of the Laws of Agricultural Science, 60

    The Laws of Modern Agriculture, 60

    Law of diminishing returns, 60
    Equilibrium, 60
    Adaptation, 60
    Compensation and cancellation, 60
    Relativity, 61
    Law of minimum, 61

    All Laws Are Meaningless, 62
    A Critical Look at Liebig's Law of Minimum, 65
    Where Specialized Research Has Gone Wrong, 68
    Critique of the Inductive and Deductive Methods, 70
    High-Yield Theory Is Full of Holes, 73

    A Model of Harvest Yields, 75
    A Look at Photosynthesis, 78
    Look Beyond the Immediate Reality, 83
    Original Factors Are Most Important, 84
    No Understanding of Causal Relationships, 86

3. The Theory of Natural Farming, 91

1. The Relative Merits of Natural Farming and Scientific Agriculture, 93

    Two Ways of Natural Farming, 93

    Mahayana Natural Farming, 93
    Hinayana Natural Farming, 93
    Scientific Farming, 93
    The Three Ways of Farming Compared, 94

    Scientific Agriculture: Farming Without Nature, 96

    1. Cases where scientific farming excels, 97
    2. Cases where both ways of farming are equally effective, 97

    The Entanglement of Natural and Scientific Farming, 99

2. The Four Principles of Natural Farming, 102

    No Cultivation, 103

    Plowing Ruins the Soil, 103
    The Soil Works Itself, 104

    No Fertilizer, 106

    Crops Depend on the Soil, 106
    Are Fertilizers Really Necessary?, 106

The Countless Evils of Fertilizer, 107
Why the Absence of No-Fertilizer Tests?, 109
Take a Good Look at Nature, 110
Fertilizer Was Never Necessary to Begin With, 111

No Weeding, 112

Is There Such a Thing as a Weed?, 112
Grasses Enrich the Soil, 113
A Cover of Grass Is Beneficial, 114

No Pesticides, 115

Insect Pests Do Not Exist, 115
Pollution by New Pesticides, 115
The Root Cause of Pine Rot, 117

3. How Should Nature Be Perceived?, 119

Seeing Nature as Wholistic, 119

Examining the Parts Never Gives a Complete Picture, 119
Become One with Nature, 120
Imperfect Human Knowledge Falls Short of Natural Perfection, 121

Do Not Look at Things Relatively, 122
Take a Perspective That Transcends Time and Space, 123
Do Not Be Led Astray by Circumstance, 125
Be Free of Cravings and Desires, 126
No Plan Is the Best Plan, 126

4. Natural Farming for a New Age, 128

At the Vanguard of Modern Farming, 128
Natural Livestock Farming, 129

The Abuses of Modern Livestock Farming, 129
Natural Grazing Is the Ideal, 129
Livestock Farming in the Search for Truth, 131

Natural Farming—In Pursuit of Nature, 133
The Only Future for Man, 133

4. The Practice of Natural Farming, 145

1. Starting a Natural Farm, 147

Keep a Natural Protected Wood, 147

Growing a Wood Preserve, 149
Shelterbelts, 149

Setting Up an Orchard, 149
Starting a Garden, 150

The Non-Integrated Garden, 151

8

Creating a Rice Paddy, 152

    Traditional Paddy Preparation, 152

Crop Rotation, 153

    Rice/Barley Cropping, 154
    Upland Rice, 154
    Minor Grains, 158
    Vegetables, 158
    Fruit Trees and Crop Rotation, 158

2. Rice and Winter Grain, 160

The Course of Rice Cultivation in Japan, 160

    Changes in Rice Cultivation Methods, 160

Barley and Wheat Cultivation, 162

    Natural Barley/Wheat Cropping, 163

        1. Tillage, ridging, and drilling, 163
        2. Light-tillage, low-ridge or level-row cultivation, 163
        3. No-tillage, direct-seeding cultivation, 164

Early Experiences with Rice Cultivation, 166
Second Thoughts on Post-Season Rice Cultivation, 168
First Steps Toward Natural Rice Farming, 171

       Natural Seeding, 171
       Natural Direct Seeding, 173

Early Attempts at Direct-Seeding, No-Tillage Rice/Barley Succession, 174

    First trial: Direct seeding of rice between barley, 174
    Second trial: Direct-seeding rice/barley succession, 174
    Third trial: Direct-seeding, no-tillage rice/barley succession, 176

Natural Rice and Barley/Wheat Cropping, 176
Direct-Seeding, No-Tillage Barley/Rice Succession with Green Manure
    Cover, 177

    Cultivation Method, 177
    Farmwork, 178

        1. Digging drainage channels, 178
        2. Harvesting, threshing, and cleaning the rice, 178
        3. Seeding clover, barley, and rice, 178
        4. Fertilization, 180
        5. Straw mulching, 181
        6. Harvesting and threshing barley, 182
        7. Irrigation and drainage, 182
        8. Disease and pest "control", 183

High-Yield Cultivation of Rice and Barley, 184

    The Ideal Form of a Rice Plant, 184
    Analysis of the Ideal Form, 186
    Ideal Shape of Rice, 187
    Blueprint for the Natural Cultivation of Ideal Rice, 188
    The Meaning and Limits of High Yields, 189

3. Fruit Trees, 193

Establishing an Orchard, 193

Natural Seedlings and Grafted Nursery Stock, 194
Orchard Management, 194

1. Correcting the tree form, 194
2. Weeds, 195
3. Terracing, 195

A Natural Three-Dimensional Orchard, 196
Building Up Orchard Earth without Fertilizers, 196

Why I Use a Ground Cover, 196
Ladino Clover, Alfalfa, and Acacia, 198
Features of Ladino Clover, 198
Seeding Ladino Clover, 198
Managing Ladino Clover, 199
Alfalfa for Arid Land, 199
Morishima Acacia, 199
Acacia Protects Natural Predators, 200
Some Basics on Setting Up a Ground Cover, 201
Soil Management, 201

Disease and Insect Control, 203

Arrowhead Scale, 204
Mites, 204
Cottony-Cushion Scale, 205
Red Wax Scale, 205
Other Insect Pests, 206
Mediterranean Fruit Fly and Codling Moth, 206

The Argument against Pruning, 207

No Basic Method, 207
Misconceptions about the Natural Form, 209
Is Pruning Really Necessary?, 210

The Natural Form of a Fruit Tree, 212

Examples of Natural Forms, 215
Attaining the Natural Form, 216
Natural Form in Fruit Tree Cultivation, 216
Problems with the Natural Form, 218

Conclusion, 220

4. Vegetables, 221

Natural Rotation of Vegetables, 221
Semi-Wild Cultivation of Vegetables, 222

A Natural Way of Growing Garden Vegetables, 222
Scattering Seed on Unused Land, 223
Things to Watch Out for, 225

Disease and Pest Resistance, 225

Resistances of Vegetables to Disease and Insects, 227
Minimal Use of Pesticides, 227

10

5. The Road Man Must Follow, 229

1. The Natural Order, 231

   Microbes as Scavengers, 233
   Pesticides in the Biosystem, 236
   Leave Nature Alone, 237

2. Natural Farming and a Natural Diet, 239

   What is Diet?, 239
   Tasty Rice, 242
   Getting a Natural Diet, 244

   Plants and Animals Live in Accordance with the Seasons, 244
   Eating with the Seasons, 247

   The Nature of Food, 251

   Color, 251
   Flavor, 252
   The Staff of Life, 255

   Summing Up Natural Diet, 257

   The Diet of Non-Discrimination, 258
   The Diet of Principle, 258
   The Diet of the Sick, 259
   Conclusion, 260

3. Farming for All, 261

   Creating True People, 261
   The Road Back to Farming, 262
   Enough Land for All, 264
   Running a Farm, 266

Epilogue, 270

Glossary of Japanese Words, 275
Translator's Note, 277
Index, 279

Preface

Natural farming is based on a nature free of human meddling and intervention. It strives to restore nature from the destruction wrought by human knowledge and action, and to resurrect a humanity divorced from God.

While still a youth, a certain turn of events set me out on the proud and lonely road back to nature. With sadness, though, I learned that one person cannot live alone. One either lives in association with people or in communion with nature. I found also, to my despair, that people were no longer truly human, and nature no longer truly natural. The noble road that rises above the world of relativity was too steep for me.

These writings are the record of one farmer who for fifty years has wandered about in search of nature. I have traveled a long way, yet as night falls, the road that remains ahead stretches far off in the distance.

Of course, in a sense, natural farming will never be perfected. It will not see general application in its true form, and will serve only as a brake to slow the mad onslaught of scientific agriculture.

Ever since I began proposing a way of farming in step with nature, I have sought to demonstrate the validity of five major principles: no tillage, no fertilizer, no pesticides, no weeding, and no pruning. During the many years that have elapsed since, I have never once doubted the possibilities of a natural way of farming that renounces all human knowledge and intervention. To the scientist convinced that nature can be understood and used through human intellect and action, natural farming is a special case and has no universality. Yet these basic principles apply everywhere.

The trees and grasses release seeds that fall to the ground, there to germinate and grow into new plants. The seeds sown by nature are not so weak as to grow only in plowed fields. Plants have always grown by direct seeding, without tillage. The soil in the fields is worked by small animals and roots, and enriched by green manure plants.

Only over the last fifty years or so have chemical fertilizers became thought of as indispensable. True, the ancient practice of using manure and compost does help speed crop growth, but this also depletes the land from which the organic material in the compost is taken.

Even organic farming, which everyone is making such a big fuss over lately, is just another type of scientific farming. A lot of trouble is taken to move organic materials first here then there, to process and treat. But any gains to be had from all this activity are local and temporal gains. In fact, when examined from a broader perspective, many such efforts to protect the natural ecology are actually destructive.

Although a thousand diseases attack plants in the fields and forests, nature strikes a balance; there never was any need for pesticides. Man grew confused when he identified these diseases as insect damage; he created with his own hands the need for labor and toil.

Man tries also to control weeds, but nature does not arbitrarily call one grass a weed and try to eradicate it. Nor does a fruit tree always grow more vigorously and bear more fruit when pruned. A tree grows best in its natural habit; the branches do not tangle, sunlight falls on every leaf, and the tree bears fruit each year, not only in alternate years.

Many people are worried today over the drying out of arable lands and loss of vegetation throughout the world, but there is no doubting that human civilization and the misguided methods of crop cultivation that arose from man's arrogance are largely responsible for this global plight.

Overgrazing by large animal herds kept by nomadic peoples has reduced the variety of vegetation, denuding the land. Agricultural societies too, with the shift to modern agriculture and its heavy reliance on petroleum-based chemicals, have had to confront the problem of rapid debilitation of the land.

Once we accept that nature has been harmed by human knowledge and action, and renounce these instruments of chaos and destruction, nature will recover its ability to nurture all forms of life. In a sense, my path to natural farming is a first step toward the restoration of nature.

That natural farming has yet to gain wide acceptance shows just how mortally nature has been afflicted by man's tampering and the extent to which the human spirit has been ravaged and ruined. All of which makes the mission of natural farming that much more critical.

I have begun thinking that the natural farming experience may be of some help, however small, in revegetating the world and stabilizing food supply. Although some will call the idea outlandish, I propose that the seeds of certain plants be sown over deserts in clay pellets to help green these barren lands.

These pellets can be prepared by first mixing the seeds of green manure trees —such as Morishima acacia—that grow in areas with an annual rainfall of less than 2 inches, and the seeds of clover, alfalfa, bur clover, and other types of green manure, with grain and vegetable seeds. The mixture of seeds is coated first with a layer of soil, then one of clay, to form microbe-containing clay pellets. These finished pellets could then be scattered by hand over the deserts and savannahs.

Once scattered, the seeds within the hard clay pellets will not sprout until rain has fallen and conditions are just right for germination. Nor will they be eaten by mice and birds. A year later, several of the plants will survive, giving a clue as to what is suited to the climate and land. In certain countries to the south, there are reported to be plants that grow on rock and trees that store water. Anything will do, as long as we get the deserts blanketed rapidly with a green cover of grass. This will bring back the rains.

While standing in an American desert, I suddenly realized that rain does not fall from the heavens; it issues forth from the ground. Deserts do not form because there is no rain; rather, rain ceases to fall because the vegetation has disappeared. Building a dam in the desert is an attempt to treat the symptoms of the disease, but is not a strategy for increasing rainfall. First we have to learn how to restore the ancient forests.

But we do not have time to launch a scientific study to determine why the deserts are spreading in the first place. Even were we to try, we would find that no matter how far back into the past we go in search of causes, these causes are preceded by other causes in an endless chain of interwoven events and factors that is beyond man's powers of comprehension. Suppose that man were able in this way to learn which plant had been the first to die off in a land turned to desert. He would still not know enough to decide whether to begin by planting the first type of vegetation to disappear or the last to survive. The reason is simple: in nature, there is no cause and effect.

Science rarely looks to microorganisms for an understanding of large causal relationships. True, the perishing of vegetation may have triggered a drought, but the plants may have died as a result of the action of some microorganism. However, botanists are not to be bothered with microorganisms as these lie outside of their field of interest. We've gathered together such a diverse collection of specialists that we've lost sight of both the starting line and the finish line. That is why I believe that the only effective approach we can take to revegetating barren land is to leave things largely up to nature.

One gram of soil on my farm contains about 100 million nitrogen-fixing bacteria and other soil-enriching microbes. I feel that soil enclosing seeds and these microorganisms could be the spark that restores the deserts.

I have created, together with the insects in my fields, a new strain of rice I call "Happy Hill." This is a hardy strain with the blood of wild variants in it, yet it is also one of the highest yielding strains of rice in the world. If a single head of Happy Hill were sent across the sea to countries where food is scarce and there sown over a ten-square-yard area, a single grain would yield 5,000 grains in one year's time. There would be grain enough to sow a half-acre the following year, fifty acres two years hence, and 7,000 acres in the fourth year. This could become the seed rice for an entire nation. This handful of grain could open up the road to independence for a starving people.

But this seed rice must be delivered as soon as possible. Even one person can begin. I could be no happier than if my humble experience with natural farming were to be used toward this end.

My greatest fear today is that of nature being made the plaything of the human intellect. There is also the danger that man will attempt to protect nature through the medium of human knowledge, without noticing that nature can be restored only by abandoning our preoccupation with knowledge and action that has driven it to the wall.

All begins by relinquishing human knowledge.

Although perhaps just the empty dream of a farmer who has sought in vain to return to nature and the side of God, I wish to become the sower of seed. Nothing would give me more joy than to meet others of the same mind.

Introduction

Anyone Can Be a Quarter-Acre Farmer

In this hilltop orchard overlooking the Inland Sea stand several mud-walled huts. Here, young people from the cities—some from other lands—live a crude, simple life growing crops. They live self-sufficiently, without electricity or running water, on a diet of brown rice and vegetables. These young fugitives, disaffected with the cities or religion, tread through my fields clad only in a loincloth. The search for the bluebird of happiness brings them to my farm in one corner of Iyo-shi in Ehime Prefecture, where they learn how to become quarter-acre farmers.

Chickens run free through the orchard, and semi-wild vegetables grow in the clover among the trees.

In the paddy fields spread out below on the Dogo Plain, one no longer sees the pastoral green of barley and the blossoms of rape and clover from another age. Instead, desolate fields lie fallow, the crumbling bundles of straw portraying the chaos of modern farming practices and the confusion in the hearts of farmers.

Only my field lies covered in the fresh green of winter grain*. This field has not been plowed or turned in over thirty years. Nor have I applied chemical fertilizers or prepared compost, or sprayed pesticides or other chemicals. I practice what I call "do-nothing" farming here, yet each year I harvest close to 22 bushels (1,300 pounds) of winter grain and 22 bushels of rice per quarter-acre. My goal is to eventually take in 33 bushels per quarter-acre.

Growing grain in this way is very easy and straightforward. I simply broadcast clover and winter grain over the ripening heads of rice before the fall harvest. Later, I harvest the rice while treading on the young shoots of winter grain. After leaving the rice to dry for three days, I thresh it then scatter the straw uncut over the entire field. If I have some chicken droppings on hand, I scatter this over the straw. Next, I form clay pellets containing seed rice and scatter the pellets over the straw before the New Year. With the winter grain growing and the rice seed sown, there is now nothing left to do until the harvesting of the winter grain. The labor of one or two people is more than enough to grow crops on a quarter acre.

In late May, while harvesting the winter grain, I notice the clover growing luxuriantly at my feet and the small shoots that have emerged from the rice seed in the clay pellets. After harvesting, drying, and threshing the winter grain, I scatter all of the straw uncut over the field. I then flood the field for four to five days to weaken the clover and give the rice shoots a chance to break through the cover of clover. In June and July, I leave the field unirrigated, and in August I run water through the drainage ditches once every week or ten days.

That is essentially all there is to the method of natural farming I shall call "direct-seeded, no-tillage, winter grain/rice succession in a clover cover."

---

*Barley or wheat. Barley cultivation is predominant in Japan, but most of what I say about barley in this book applies equally well to wheat.

Were I to say that all my method of farming boils down to is the symbiosis of rice and barley or wheat in clover, I would probably be reproached: "If that's all there is to growing rice, then farmers wouldn't be out there working so hard in their fields." Yet, that *is* all there is to it. Indeed, with this method I have consistently gotten better-than-average yields. Such being the case, the only conclusion possible is that there must be something drastically wrong with farming practices that require so much unnecessary labor.

Scientists are always saying, "Let's try this, let's try that." Agriculture becomes swept up in all of this fiddling around; new methods requiring additional expenditures and effort by farmers are constantly introduced, along with new pesticides· and fertilizers. As for me, I have taken the opposite tack. I eliminate unnecessary practices, expenditures, and labor by telling myself, "I don't need to do this, I don't need to do that." After thirty years at it, I have managed to reduce my labor to essentially just sowing seed and spreading straw. Human effort is unnecessary because nature, not man, grows the rice and wheat.

If you stop and think about it, every time someone says "this is useful," "that has value," or "one ought to do such-and-such," it is because man has created the preconditions that give this whatever-it-is its value. We create situations in which, without something we never needed in the first place, we are lost. And to get ourselves out of such a predicament, we make what appear to be new discoveries, which we then herald as progress.

Flood a field with water and stir it up with a plow and the ground will set, becoming as hard as plaster. If the soil dies and hardens, then it must be plowed each year to soften it. All we are doing is creating the conditions that make a plow useful, then rejoicing at the utility of our tool. No plant on the face of the earth is so weak as to germinate only in plowed soil. Man has no need to plow and turn the earth, for microorganisms and small animals act as nature's tillers.

By killing the soil with plow and chemical fertilizer, and rotting the roots through prolonged summer flooding, farmers create weak, diseased rice plants that require the nutritive boost of chemical fertilizers and the protection of pesticides. Healthy rice plants have no need for the plow or chemicals. And compost does not have to be prepared if rice straw is applied to the fields half a year before the rice is sown.

Soil enriches itself year in and year out without man having to lift a finger. On the other hand, pesticides ruin the soil and create a pollution problem. Shrines in Japanese villages are often surrounded by a grove of tall trees. These trees were not grown with the aid of nutrition science, nor are they protected by plant ecology.

Saved from the axe and saw by the shrine deity, they grow into large trees of their own accord.

Properly speaking, nature is neither living nor dead. Nor is it small or large, weak or strong, thriving or feeble. It is those who believe only in science who call an insect either a pest or a natural enemy and cry out that nature is a violent world of relativity and contradiction in which the strong feed on the weak. Notions of right and wrong, good and bad, are alien to nature. These are only distinctions invented by man. Nature maintained a great harmony without such notions, and brought forth the grasses and trees without the "helping" hand of man.

The living and wholistic biosystem that is nature cannot be dissected or resolved into its parts. Once broken down, it dies. Or rather, those who break off a piece of nature lay hold of something that is dead, and unaware that what they are examining is no longer what they think it to be, claim to understood nature. Man commits a grave error when he collects data and findings piecemeal on a dead and fragmented nature and claims to "know," "use," or "conquer" nature. Because he starts off with misconceptions about nature and takes the wrong approach to understanding it, regardless of how rational his thinking, everything winds up all wrong. We must become aware of the insignificance of human knowledge and activity, and begin by grasping their uselessness and futility.

Follow the Workings of Nature

We often speak of "producing food," but farmers do not produce the food of life. Only nature has the power to produce something from nothing. Farmers merely assist nature.

Modern agriculture is just another processing industry that uses oil energy in the form of fertilizers, pesticides, and machinery to manufacture synthetic food products which are poor imitations of natural food. The farmer today has become a hired hand of industrialized society. He tries without success to make money at farming with synthetic chemicals, a feat that would tax even the powers of the Thousand-Handed Goddess of Mercy. It is no surprise then that he is spinning around like a top.

Natural farming, the true and original form of agriculture, is the methodless method of nature, the unmoving way of Bodhidharma. Although appearing fragile and vulnerable, it is potent for it brings victory unfought; it is a Buddhist way of farming that is boundless and yielding, and leaves the soil, the grasses, and the insects to themselves.

As I walk through the paddy field, spiders and frogs scramble about, locusts jump up, and droves of dragonflies hover overhead. Whenever a large outbreak of leafhoppers occurs, the spiders multiply too, without fail. Although the yield of this field varies from year to year, there are generally about 250 heads of grain per square yard. With an average of 200 grains per head, this gives a harvest of some 33 bushels for every quarter-acre. Those who see the sturdy heads of rice rising from the field marvel at the strength and vigor of the plants and their large yields. No matter that there are insect pests here. As long as their natural enemies are also present, a natural balance asserts itself.

Since it is founded upon principles derived from a fundamental view of nature, natural farming remains current and applicable in any age: Although ancient, it is also forever new. Of course, such a way of natural farming must be able to weather the criticism of science. The question of greatest concern is whether this "green philosophy" and way of farming has the strength to criticize science and lead man on the road back to nature.

18

Fig. A. Rice cultivation by natural farming.

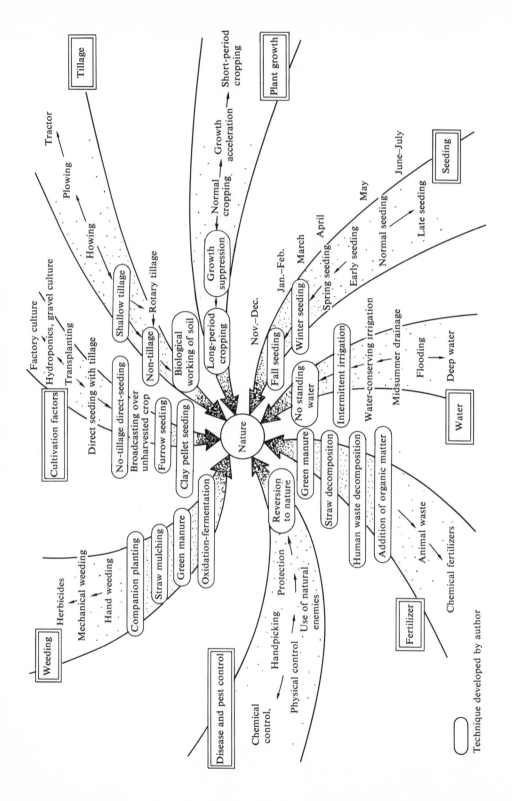

Technique developed by author

Fig. B. Rice cultivation by scientific farming.

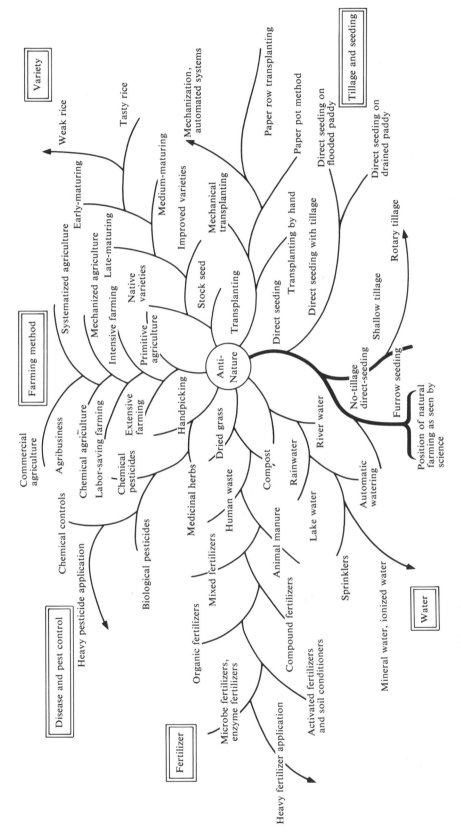

With the growing popularity of natural foods lately, I thought that natural farming too would be studied at last by scientists and receive the attention it is due. Alas, I was wrong. Although some research is being conducted on natural farming, most of it remains strictly within the scope of scientific agriculture as practiced to date. This research adopts the basic framework of natural farming, but makes not the slightest reduction in the use of chemical fertilizers and pesticides; even the equipment used has gotten larger and larger.

Why do things turn out this way? Because scientists believe that, by adding technical know-how to natural farming, which already reaps over 22 bushels of rice per quarter acre, they will develop an even better method of cultivation and

Fig. C.   Toward a natural way of farming.

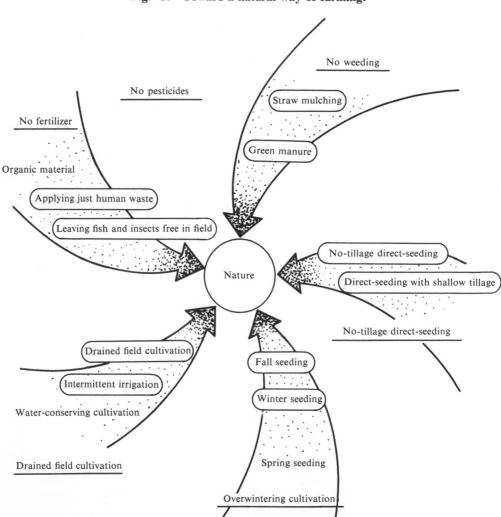

higher yields. Although such reasoning appears to make sense, one cannot ignore the basic contradiction it entails. Until the day that people understand what is meant by "doing nothing"—the ultimate goal of natural farming, they will not relinquish their faith in the omnipotence of science.

When we compare natural farming and scientific farming graphically, we can right away appreciate the differences between the two methods. The objective of natural farming is non-action and a return to nature; it is centrifugal and convergent. On the other hand, scientific farming breaks away from nature with the expansion of human wants and desires; it is centripetal and divergent. Because this outward expansion cannot be stopped, scientific farming is doomed to extinction. The addition of new technology only makes it more complex and diversified, generating ever-increasing expense and labor. In contrast, not only is natural farming simple, it is also economical and labor-saving.

Fig. D.   The direction taken by scientific agriculture.

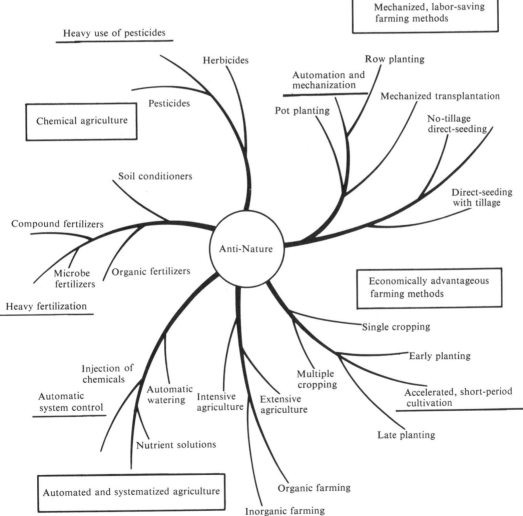

Why is it that, even when the advantages are so clear and irrefutable, man is unable to walk away from scientific agriculture? People think, no doubt, that "doing nothing" is defeatist and hurts production and productivity. Yet, does natural farming hurt productivity? Far from it. In fact, if we base our figures on the efficiency of energy used in production, natural farming turns out to be the most productive method of farming there is.

Natural farming produces 130 pounds of rice, or 200,000 kilocalories of energy, per man-day of labor, without the input of any outside materials. This is about 100 times the daily intake of 2,000 kilocalories by a farmer on a natural diet. Ten times as much energy was expended in traditional farming, which used horses and oxen to plow the fields. The energy input in calories was doubled again with the advent of small-scale mechanization, and doubled yet another time with the shift to large-scale mechanization. This progression has given us the energy-intensive agricultural methods of today.

The claim is often made that mechanization has increased the efficiency of work, but farmers must use the extra hours away from their fields to earn outside income to help pay for their equipment. All they have done is exchange their work in the fields for a job in some company; they have traded the joy of working out in the open fields for dreary hours of labor shut up inside a factory.

People believe that modern agriculture can both improve productivity and increase yields. What a misconception. The truth of the matter is that the yields provided by scientific farming are smaller than the yields attainable with the full powers of nature. High-yield practices and scientific methods of increasing production are thought to have given us increased yields that exceed the natural productivity of the land, but this is not so. These are merely endeavors by man to artificially restore full productivity after he has hamstrung nature so that it cannot exercise its full power. Man creates adverse conditions, then rejoices later at his "conquest" of nature. High-yield technologies amount to no more than glorified attempts to stave off reductions in productivity.

Nor is science a match for nature in terms of the quality of the food it helps to create. Ever since man deluded himself into thinking that nature can be understood by being broken down and analyzed, scientific farming has produced artificial, deformed food. Modern agriculture has created nothing from nature. Rather, by making quantitative and qualitative changes in certain aspects of nature, it has managed only to fabricate crude, expensive, and synthetic food products that further alienate man from nature.

Humanity has left the bosom of nature and only recently begun to view with growing alarm its position as orphan of the universe. Yet, even were he to try returning to nature, man would find that he no longer knows what nature is, and that, moreover, he has destroyed and forever lost the nature he seeks to return to.

Scientists envision domed cities of the future in which enormous heaters, air conditioners, and ventilators will provide comfortable living conditions throughout the year. They dream of building underground cities and colonies on the seafloor. But the city dweller is dying; he has forgotten the bright rays of the sun, the green fields, the plants and animals, and the pleasant sensation of a gentle breeze on the skin. Man can live a true life only with nature.

Natural farming is a Buddhist way of farming that originates in the philosophy of "Mu," or nothingness, and returns to a "do-nothing" nature. The young people living in my orchard carry with them the hope of someday resolving the great problems of our world that cannot be solved by science and reason. Mere dreams perhaps, but these hold the key to the future.

Ailing Agriculture in an Ailing Age

1.   Man Cannot Know Nature

Man prides himself on being the only creature on earth with the ability to think. He claims to know himself and the natural world, and believes he can use nature as he pleases. He is convinced, moreover, that intelligence is strength, that anything he desires is within his reach.

As he has forged ahead, making new advances in the natural sciences and dizzily expanding his materialistic culture, man has grown estranged from nature and ended by building a civilization all his own, like a wayward child rebelling against its mother.

But all his vast cities and frenetic cultural and economic activities have brought him are empty, dehumanized pleasures and the destruction of his living environment through the abusive exploitation of nature.

Harsh retribution for straying from nature and plundering its riches have begun to appear in the form of depleted natural resources and food crises, throwing a dark shadow on man's future. Having finally grown aware of the gravity of the situation, man has begun to think seriously about what should be done, but unless he is willing to undertake the most fundamental self-reflection he will be unable to steer away from a path of certain destruction.

Alienated from nature, human existence becomes a void, the wellspring of life and spiritual growth gone utterly dry. Man grows ever more ill and weary in the midst of his curious civilization that is but a struggle over a tiny bit of time and space.

Leave Nature Alone

Man has always deluded himself into thinking that he knows nature and is free to use it as he wishes to build his civilization. But nature cannot be explained or expanded upon. As an organic whole, it is not subject to man's classifications; nor does it tolerate dissection and analysis. Once broken down, nature cannot be returned to its original state. All that remains is an empty skeleton devoid of the true essence of living nature. This skeletal image only serves to confuse man and lead him further astray. Nor is scientific reasoning of any avail in helping man understand nature and add to its creations.

Nature as perceived by man through discriminating knowledge is a falsehood. Man can never truly know even a single leaf, even a single handful of earth. Unable to fully comprehend plant life and soil, he sees these only through the filter of human intellect.

Although he may seek to return to the bosom of nature or use it to his advantage, he only touches one tiny part of nature—a dead portion at that—and has no affinity with the main body of living nature. He is, in effect, merely toying with delusions.

Man is but an arrogant fool who vainly believes that he knows all of nature and can achieve anything he sets his mind to. Seeing neither the logic nor order inherent in nature, he has selfishly appropriated it to his own ends and destroyed it. The world today is in such a sad state because man has not felt compelled to reflect upon the dangers of his high-handed ways.

The earth is an organically interwoven community of plants, animals, and microorganisms. When seen through man's eyes, it appears either as a model of the strong consuming the weak or of coexistence and mutual benefit. Yet there are food chains and cycles of matter; there is endless transformation without birth or death. This flux of matter and the cycles in the biosphere can be perceived only through direct intuition, yet our unswerving faith in the omnipotence of science has led us to analyze and study these phenomena, raining down destruction upon the world of living things and throwing nature as we see it into disarray.

A case in point is the application of toxic pesticides to apple trees and hothouse strawberries. This kills off pollenating insects such as bees and gadflies, forcing man to collect the pollen himself and artificially pollenate each of the blossoms. Although he cannot even hope to replace the myriad activities of the innumerable plants, animals, and microorganisms in nature, man goes out of his way to block their activities, then studies each of these functions carefully and attempts to find substitutes. What a ridiculous waste of effort.

Consider the case of the scientist who studies mice and develops a rodenticide. He does so without understanding why mice flourished in the first place. He simply decides that killing them is a good idea without first determining whether the mice multiplied as the result of a breakdown in the balance of nature, or whether they support that balance. The rodenticide is a temporary expedient that answers only the needs of a given time and place; it is not a responsible action in keeping with the true cycles of nature. Man cannot possibly replace all the functions of plants and animals on this earth through scientific analysis and human knowledge. While unable to fully grasp the totality of these interrelationships, any rash endeavor such as the selective extermination or raising of a species only serves to upset the balance and order of nature.

Even the replanting of mountain forests may be seen as destructive. Trees are logged for their value as lumber, and species of economic value to man, such as pine and cedar, are planted in large number. We even go so far as to call this "forestry conservation." However, altering the tree cover on a mountain produces changes in the characteristics of the forest soil, which in turn affects the plants and animals that inhabit the forest. Qualitative changes also take place in the air and temperature of the forest, causing subtle changes in weather and affecting the microbial world.

No matter how closely one looks, there is no limit to the complexity and detail with which nature interacts to effect constant, organic change.

When a section of the forest is clear-cut and cedar trees planted, for example, there no longer is enough food for small birds. These disappear, allowing long-horned beetles to flourish. The beetles are vectors for nematodes, which attack red pines and feed on parasitic *Botrytis* fungi in the trunks of the pine trees. The pines fall victim to the *Botrytis* fungi because they are weakened by the disappearance of the edible *matsutake* fungus that lives symbiotically on the roots of red pines. This beneficial fungus has died off as a result of an increase in the harmful *Botrytis* fungus in the soil, which is itself a consequence of the acidity of the soil. The high soil acidity is the result of atmospheric pollution and acid rain, and so on and so forth. This backward regression from effect to prior cause continues in an unending chain that leaves one wondering what the true cause is.

When the pines die, thickets of bamboo grass rise up. Mice feed on the abundant bamboo grass berries and multiply. The mice attack the cedar saplings, so man applies a rodenticide. But as the mice vanish, the weasels and snakes that feed on them decrease in number. To protect the weasels, man then begins to raise mice to restore the rodent population. Is not this the stuff of crazed dreams?

Toxic chemicals are applied at least eight times a year on Japanese rice fields. Is it not odd then that hardly any agricultural scientists have bothered to investigate why the amount of insect damage in these fields remains largely the same as in fields where no pesticides are used? The first application of pesticide does not kill off the swarms of rice leafhoppers, but the tens of thousands of young spiders on each square yard of land simply vanish, leaving few survivors, and the droves of fireflies that fly up from the stands of grass disappear at once. The second application kills off the chalcid flies, which are important natural predators, and leaves victim dragonfly larvae, tadpoles, and loaches. Just one look at this slaughter would suffice to show the insanity of the blanket application of pesticides.

No matter how hard he tries, man can never rule over nature. What he can do is serve nature, which means living in accordance with its laws.

The "Do-Nothing" Movement

The age of aggressive expansion in our materialistic culture is at an end, and a new "do-nothing" age of consolidation and convergence has arrived. Man must hurry to establish a new way of life and a spiritual culture founded on communion with nature, lest he grow ever more weak and feeble while running around in a frenzy of wasted effort and confusion.

When he turns back to nature and seeks to learn the essence of a tree or a blade of grass, man will have no need for human knowledge. It will be enough to live in concert with nature, free of plans, designs, and effort. One can break free of the false image of nature conceived by the human intellect only by becoming detached and earnestly begging for a return to the absolute realm of nature. No, not even entreaty and supplication are necessary; it is enough only to farm the earth free of concern and desire.

To achieve a humanity and a society where nothing needs to be done, man must look back over everything he has done and rid himself one by one of the false visions and concepts that permeate him and his society. This is what the "do-nothing" movement is all about.

Natural farming can be seen as one branch of this movement. Human knowledge and effort expand, spread, and grow increasingly complex and wasteful without limit. We need to halt this expansion, to converge, simplify, and reduce our knowledge and effort. This is in keeping with the laws of nature. Natural farming is more than just a revolution in agricultural techniques. It is the practical foundation of a spiritual movement, of a revolution to change the way man lives.

2. The Breakdown of Japanese Agriculture

Life in the Farming Villages of the Past

In earlier days, Japanese peasants were a poor and downtrodden lot. Forever oppressed by those in power, they occupied the lowest rung on the social ladder. Where did they find the strength to endure their poverty and what did they depend on to live?

The farmers who lived quietly in a secluded inland glen, on a solitary island in the southern seas, or in a desolate northern region of deep snows were self-supporting and independent; they lived a proud, happy, noble life in the great outdoors. People born in remote areas who lived out poor lives and died anonymously were able to subsist in a world cut off from the rest of mankind without discontent or anxiety because, though they appeared alone, they were not. They were creatures of nature, and being close to God (nature incarnate), experienced the daily joy and pride of tending the gardens of God. They went out to work in the fields at sunrise and returned home to rest at sunset, living each day well, one day being as wide and infinite as the universe and yet just one small frame in the unending flow of existence. Theirs was a farming way of life, in the midst of nature, which violated nothing and was not itself violated.

Farmers are bound to take offense when the clever smart alecks who left the village and made their way in the world come back, saying "sir, sir" with false humility, then, when you least expect it, telling you in effect to "go to hell." Although farmers have no need for business cards, on occasion they have been misers too mean to part with a single penny, and at other times, millionaires without the slightest interest in fabulous riches. Peasant villages were lonely, out-of-the-way places inhabited by indigent farmers, yet were also home to recluses who lived in a world of the sublime. People in the small, humble villages of which Lao-tzu spoke were unaware that the Great Way of man lay in living independently and self-sufficiently, yet they knew this in their hearts. These were the farmers of old.

What a tragedy it would be to think of these farmers as fools who know, yet are unaware. To the remark that "any fool can farm," farmers should reply, "a fool cannot be a true farmer." There is no need for philosophy in the farming village. It is the urban intellectual who ponders human existence, who goes in search of truth and questions the purpose of life.

The farmer does not wrestle with the questions of why man arose on the face of the earth and how he should live. Why is it that he never learned to question his existence? Life was never so empty and void as to bring him to contemplate the purpose of human existence; there was no seed of uncertainty to lead him astray.

With their intuitive understanding of life and death, these farmers were free of anguish and grief; they had no need for learning. They joked that agonizing over life and death, and wandering through ideological thickets in search of truth were the pastimes of idle city youth. Farmers preferred to live common lives, without knowledge or learning. There was no time for philosophizing. Nor was there any need. This does not mean that the farming village was without a philosophy. On the contrary, it had a very important philosophy. This was embodied in the principle that "philosophy is unnecessary." The farming village was above all a society of philosophers without a need for philosophy. It was none other than the philosophy of Mu, or nothingness—which teaches that nothing is of any avail, that gave the farmer his enduring strength.

Disappearance of the Village Philosophy

Not that long ago one could still hear the woodsman sing a woodcutter's song as he sawed down a tree. During transplanting, singing voices rolled over the paddy fields, and the sound of drums surged through the village after the fall harvest. Nor was it that long ago that people used pack animals to carry goods.

These scenes have changed drastically over the past twenty years or so. In the mountains, instead of the rasping of hand saws, we now hear the angry snarl of chain saws. We see mechanical plows and transplanters racing over the fields. Vegetables today are grown in vinyl houses ranged in neat rows like factories. The fields are automatically sprayed with fertilizers and pesticides. Because all of the farmer's work has been mechanized and systematized, the farming village has lost its human touch. Singing voices are no longer heard. Everyone sits instead before the TV set, listening to traditional country songs and reminiscing over the past.

We have fallen from a true way of life to one that is false and untrue. People rush about in a frenzy to shorten time and widen space, and in so doing lose both.

The farmer may have thought at first that modern developments would make his job easier. Well, it freed him from the land and now he works harder than ever at other jobs, wearing away his body and mind.

The chain saw was developed because someone decided that a tree had to be cut faster. Rather than making things easier for the farmer, the mechanized transplantation of rice has sent him running off to find other work.

The disappearance of the sunken hearth from farming homes has extinguished the light of ancient farming village culture. Fireside discussions have vanished, and with them, the village philosophy.

High Growth and the Farming Population after World War II

No country has experienced such a sudden and dramatic transformation following World War II as Japan, which rose rapidly from the ruins of war to become a major economic power. As this was going on, its farming and fishing populations —the seedbed of the Japanese people—fell from fifty percent of the overall population at the end of the war to less than twenty percent today. Without the help of the dexterous, hard-working farmer, the skyscrapers, highways, and subways of the big cities would never have materialized. Japan owes its current prosperity to the labor it appropriated from the farming population and placed at the service of urban civilization.

Japan's rapid growth following the war is generally attributed to good fortune and wise leadership. However, the farmer draws a different interpretation. Changes in the self-image of the farming population led to the adoption of new agricultural methods. As farming became less labor-intensive, surplus manpower poured out of the countryside into the towns and cities, bringing prosperity to the urban civilization. But far from being a blessing, this prosperity has made things harder on the farmer. In effect, he tightened the noose about his own neck. How did this happen?

The first step was the arrival of the motorized transport-tiller in the farming village, a major turning point in Japanese agriculture. This was rapidly followed by three-wheeled vehicles and trucks. Before one knew it, ropeways, monorails, and paved roads stretched to the furthest corners of the village, all of which completely altered the farmer's notions of time and space.

With this wave of change from labor-intensive to capital-intensive farming came the replacement of the horse-drawn plow with tillers, and later, tractors. Motorized hand sprayers were abandoned in favor of helicopter spraying, and methods of pesticide and fertilizer application underwent major revisions. Needless to say, traditional farming with draft animals was abandoned and replaced with methods involving the heavy application of chemical fertilizers and pesticides.

The rapid mechanization of agriculture lit the fires for the revival and precipitous growth of the machine industry, while the adoption of pesticides, chemical fertilizers, and petroleum-based farming materials laid the foundation for development of the chemical industry.

It was the desire by farmers to modernize, the sweeping reforms in methods of crop cultivation, that opened up the road to a new transformation of society following the destruction of the weapons industry and the industrial infrastructure during the war. What began as a movement to assure adequate food supplies in times of acute shortage grew into a drive to increase food production, the momentum of which carried over into the industrial world. This is where things stood in the mid-1950s.

The situation changed completely in the late sixties and early seventies. Stability of food supply had been achieved for the most part and the economy was overflowing with vigor. At last the visions of a modern industrial state were beginning to be realized. It was at about this time that politicians and businessmen started thinking of how to bring the large number of farmers and their land into the picture.

Once food surpluses started to arise, the farmers became a weight around the government's neck. The food control system set up to ensure an adequate food supply began to be regarded as a burden on the nation. The Fundamentals of Agriculture Act was established in 1961 to define the role and direction to be taken by Japanese agriculture. But instead of serving as a foundation for farmers, it established controls over the farmer and passed the reins of control to the financial community.

The general public started thinking that agricultural land could be put to better use in industry and housing than for food production; city dwellers even began to see farmers, who were reluctant to part with their land, as selfish monopolizers of land. Laborers and office workers joined in the effort to drive farmers off their land and taxes as high as those on housing land were levied on farmland.

The effort by farmers to raise food production appears to have backfired against them. Even though Japan's food self-sufficiency has dropped below thirty percent, farmers are unable to speak up because the people of the nation are under the illusion that the farmland reduction policy being pushed through by the government is in the interest of the consumer. Somewhere along the way, the farmer lost both his land and the freedom to choose the crops he wishes to raise. Farmers have simply gone with the flow of the times. Today, most of them lament that they can't make a decent living off farming.

Why has the farming community fallen to such a hopeless state? The experience of Japanese farmers over the past 30 years is unprecedented, and presents very grave problems for the future. Let us take a closer look at the fall of Japanese agriculture to determine exactly what happened.

How an Impoverished National Agricultural Policy Arose

When I look closely at the recent history of an agriculture that, unable to oppose the current of the times, has been made to bend and twist to the designs of the leadership, as a farmer, I cannot help feeling tremendous rage.

Behind the claim that today's farming youth is being carefully trained as agricultural specialists and model farmers lie plans to wipe out small farms and proposals for a euthanasia of farming. Underlying the spectacular programs for modernizing agriculture and increasing productivity, and the calls to expand the scale of farming operations, lies a thinly-disguised contempt for the farmer.

While the one-acre farmer was doing all he could to work his way up to three or even five acres, the policy leaders in government were saying that ten acres just was not large enough, and were running demonstration farms of 150 acres. Clearly, no matter how hard they tried to scale up their operations, farmers were pitted one against another in a form of natural selection that could only escalate into vendettas and infighting.

To the economists, who supported the doctrine of international specialization of labor, the physiocracy and stubborn insistence by farmers that their mission was to produce food were evidence of the obstinate, mule-headed farming temperament, which they despised. As for the trading companies, their basic formula for prosperity was to encourage ever more domestic and foreign food trade.

Consumers are easily won over by arguments that they "have the right to buy cheap, tasty rice." But "tasty" rice is weak rice, polluting rice grown with lots of pesticides. Such demands make things harder on the farmer, and the consumer actually ends up eating bad-tasting rice. The only one who wins out is the merchant.

People talk of "cheap rice," but it has never been the farmer who sets the price of rice or other farm produce. Nor is it the farmer who determines production costs. The price of rice nowadays is the price calculated to support the agricultural equipment manufacturers, it is the price needed for the production of new farm implements, it is the price at which fuel can be bought.

When I visited the United States in the summer of 1979, the price of rice on the

U.S. market was everywhere about 50 cents per pound—about the same as that of economy rice in Japan. Since the price of gasoline at the time was about one dollar per gallon, I was at a loss to understand the reasoning behind reports then in circulation to the effect that rice could easily be imported into Japan at one-quarter to one-third the local price. Just as incredible were reports that the surplus of rice had "left the food control system in the red" or that the scarcity of wheat had "kept the system solvent."

In natural farming, the cost of producing rice is almost the same as the cost of wheat production. Moreover, both can be produced more cheaply this way than buying imported grain. The mechanisms by which the market price of rice is set has nothing whatsoever to do with farmers. The retail price of farm produce is said to be too high in Japan, but this is because the costs of distribution are too high. Distribution costs in Japan are five times as high as in the United States and twice as high as in West Germany. One cannot help suspecting that the aim of Japan's food policy is to find the best way to line government coffers with gold. The federal assistance given per farmer is twice as high in the United States as in Japan, and three times as high in France. Japanese farmers are treated with indifference.

Today's farmers are besieged from all sides. Angry voices rise from the cities, crying, "Farmers are overprotected," "They are over-subsidized," "They're producing too much rice, putting the food control system in debt and raising our taxes."

But these are just the superficial views of people who don't see the whole picture or have any idea of the real state of affairs. I am even tempted to call these false rumors created by the gimmickry of an insanely complex society. At one time, six farming households supported one official. Today, there is reportedly one agriculture or forestry official for every full-time farmer. One wonders then if the agricultural deficits in Japan are really the fault of the farmer.

Statistics tell us that the average American farmer feeds one hundred people and the average Japanese farmer only ten, but Japanese farmers actually have a higher productivity than American farmers. It just appears the other way around because Americans farm under much better conditions than Japanese farmers.

Today's Japanese farmers are in love with money. They no longer have any time or affection for nature or their crops. All they have time for anymore is to blindly follow the figures spit out by distribution industry computers and the plans of agricultural administrators. They don't talk with the land or converse with the crops; they are interested only in money crops. They grow produce without choosing the time or place, with giving a thought to the suitability of the land or crop.

The way administrators see it, grain produced abroad and grain grown locally both have the same value. They make no distinction over whether a crop is a short-term or long-term crop. Without giving the slightest thought to the concerns of the farmer, the official instructs the farmer to grow vegetables today, fruits tomorrow, and to forget about rice. However, crop production within the natural ecosystem is no simple matter that can be resolved in an administrative bulletin. It is no wonder then that measures planned from on high are always thwarted and delayed.

When the farmer forgets the land to which he owes his existence and becomes concerned only with his own self-interest, when the consumer is no longer able to distinguish between food as the staff of life and food as merely nutrition, when the administrator looks down his nose at farmers and the industrialist scoffs at nature, then the land will answer with its death. Nature is not so kind as to forewarn a humanity so foolish as this.

What Lies Ahead for Modern Agriculture

In 1979, I boarded a plane for the first time and visited the United States. I was astounded by what I saw. I had thought that desertification and the disappearance of native peoples were stories from ancient history—in the Middle East and Africa. But I learned that the very same thing has happened repeatedly in the U.S.

Because meat is the food staple in America, agriculture is dominated by livestock farming. Grazing has destroyed the ecology of natural grasses, devastating the land. I watched this happening and could hardly believe my eyes. Land that has lost its fertility is barren of nature's strength. This accounts for the development of a modern agriculture totally reliant on petroleum energy.

The low productivity of the land drives farmers to large-scale operation. Large operations require mechanization with machinery of increasing size. This "big iron" breaks down the structure of the soil, setting up a negative cycle. Agriculture that ignores the forces of nature and relies solely on the human intellect and human effort is unprofitable. It was inevitable that these crops, produced as they are with the help of petroleum, would be transformed into a strategic commodity for securing cheap oil.

To get an idea of just how fragile commercial agriculture is with its large-scale monoculture farming on a subcontractor basis, just consider that U.S. farmers working 500 to 700 acres have smaller net incomes than Japanese farmers on 3 to 5 acres.

I realized, however, that these faults of modern farming were rooted in the basic illusions of Western philosophy that support the foundations of scientific agriculture. And I saw too that mistaken ideology had led man astray in how he lived his life and secured his essentials of food, clothing, and shelter. I noted that confusion over food had bred confusion over farming, which had destroyed nature, and I understood also that the destruction of nature had enfeebled man and thrown the world into disarray.

Is There a Future for Natural Farming?

I do not wish merely to expose and attack the current state of modern agriculture, but to point out the errors of Western thought and call for the observance of the Eastern philosophy of Mu. While recalling the self-sufficient farming practices and natural diets of the past, my desire has been to establish a natural way of farming for the future and explore the potential for its spread and adoption by others.

Yet I suppose that whether natural farming becomes the method of farming for the future depends both on a general acceptance of the thinking on which it is based and on a reversal in the existing value system. Although I will not expound here on this philosophy of Mu and its system of values, I would like to take a brief look at the agriculture of the future from the perspective of Mu.

Forty years ago, I predicted that the age of centrifugal expansion fed by the growing material desires of man, the era of rampant modern science, would soon pass and be replaced by a period of contraction and convergence as man sought to improve his spiritual life. I take it that I was wrong. Even organic farming, which has come into its own with the pollution problem, only serves as a temporary stopgap, a brief respite.

Organic farming is essentially a rehashing of the animal-based traditional farming of the past. Being part and parcel of scientific agriculture to begin with, it will be swallowed whole and assimilated by scientific agriculture.

I had hoped that the self-sufficient agriculture of the past and farming methods that try to tap into the natural ecosystem would help turn Japanese thinking around and reorient it toward natural farming—the true way of agriculture, but the current situation is almost behind hope.

Science Continues on an Unending Rampage

In today's society, man is cut off from nature and human knowledge is arbitrary. To take an example, suppose that a scientist wants to understand nature. He may begin by studying a leaf, but as his investigation progresses down to the level of molecules, atoms, and elementary particles, he loses sight of the original leaf.

Nuclear fission and fusion research is among the most advanced and dynamic fields of inquiry today, and with the development of genetic engineering, man has acquired the ability to alter life as he pleases. A self-appointed surrogate of the Creator, he has gotten hold of a magic wand, a sorcerer's staff.

And what is man likely to attempt in the field of agriculture? He probably intends to begin with the creation of curious plants by interspecific genetic recombination. It should be easy to create gigantic varieties of rice. Trees will be crossed with bamboo and eggplants will be grown on cucumber vines. It will even become possible to ripen tomatoes on trees.

By transferring genes from leguminous plants to tomato or rice, scientists will produce rhizobium-bearing tomatoes capable of fixing nitrogen from the air. Once tomatoes and rice are developed that do not require nitrogen fertilizer, farmers will no doubt jump at the chance to grow these.

Genetic engineering will most certainly be applied to insects as well. If hybrid bee-flies are created, or butterfly-dragonflies, we will no longer be able to tell whether these are beneficial insects or pests. Yet, just as the queen ant produces nothing but worker ants, man will try to create any insect or animal that is of benefit to him.

Eventually, things may progress to the point where hybrids of foxes and raccoons will be created for zoos, and we may see vegetable-like or mechanical humans created as workers. The most ridiculous products, if developed initially for the sake of medicine, let us say, will receive the plaudits of the world and win wide acceptance. A good example is the recent news, received as a godsend, that the mass production of insulin has been achieved through genetic recombination using *E. coli* genes.

The Illusions of Science and the Farmer

Today we have test-tube babies, and scientists are already envisioning a day, not that far off, when they will breed humans in culture media to create a spectrum of human races of gifted physicists, mathematicians, and whatnot. There will no longer be any need to go through the ordeal of giving birth and raising children. Children will be raised in complete incubators equipped with dispensers supplying artificial protein foods and vitamins.

No longer will food consist of unappetizing meat protein synthesized from petro-chemicals. Instead, we will enjoy delicious, inexpensive meat-like products created by crossing the genes of the soybean with the genes of a cow-pig.

Such dreams of science are so close to being achieved, I can see them as if they were already a reality. When that day does come, what will be the role of farmers then? Working the open fields under the sun may become a thing of the past. The farmer may find himself assisting the scientist as a laborer in a tightly sealed factory—perhaps even one for mass-producing bright, strong, artificial humans to eliminate the trouble of using or dealing with ordinary human beings.

To the scientist, this sort of tragedy appears as but a temporary inconvenience, a necessary sacrifice. Firm and unshaking in his conviction that, while still imperfect, someday human knowledge will be complete, that knowledge is of value as long as it is not put to the wrong use, he will probably continue to rise eagerly to the challenge of empty possibilities.

But these dreams of scientists are just mirages, nothing more than wild dancing in the hand of the Lord Buddha. Even if scientists change the living and nonliving as they please and create new life, the fruits and creations of human knowledge can never exceed the limits of the human intellect. In the eyes of nature, actions that arise from human knowledge are all futile.

All is arbitrary delusion created by the false reasoning of man in a world of relativity. Man has learned and achieved nothing. He has destroyed nature under the illusion that he controls it. He has cast himself as a plaything and crippled himself; he has brought the earth to the abyss of annihilation. Nor will it be just the farmer who follows the bidding of the scientist and lends him a hand. What a tragedy if this is what awaits the farmer of tomorrow. What a tragedy too for those who laugh at the ruin of each farmer, and those as well who merely look on.

All that remains is a last glimmer of hope that the principle dying like a buried ember in the farming village will be unearthed and revived in time to establish a natural way of farming that unites man and nature.

3.   Disappearance of a Natural Diet

Decline in the Quality of Food

It should have come as no surprise that crops grown with vast amounts of petroleum energy would suffer a decline in quality. The use of oil-based energy in agriculture has gotten to the point where one could almost talk of growing rice in the "oil patch" rather than in the "paddy."

Farming under the open skies has disappeared. Agriculture today has been degraded to the manufacture of petroleum-derived foods, and the farmer has become a seller of false goods called "nutritional food."

Ever since the farmer who had worked hand in hand with nature capitulated to the pressures of society and became a subcontractor to the oil industry, control over his livelihood has passed into the hands of the industrialist and businessman. Today it is the merchant who has the last say over the farmer's right to loss or gain, life or death.

The destruction of agriculture can be seen, for example, in the transition by farmers from the open cultivation of vegetables to hothouse horticulture. This began with the seeding and growing of melons and tomatoes in soil within hot beds or vinyl houses arranged in neat rows. The next stage was sand culture and gravel culture using sand or gravel in place of soil because these materials have fewer bacteria and are thus "cleaner." This was accompanied by a change in thinking—from forming rich soil to administering nutrients—which led to the creation and supply of nutrient solutions. The only function of the sand and gravel was to support the plant, so a simpler, more readily available material was sought. Plastic or polymer netting and containers were developed in which seeds are "planted." As these germinate and grow, the roots extend out in all directions within the plastic netting. The stem and leaves are also artificially supported, and the tightly sealed chamber in which the plants are grown is completely sterile, eliminating the chance, at first, of insect damage or blight.

Since the root absorption of nutrients dissolved in water is inefficient, the nutrient solution is sprayed on a regular basis over the entire plant. Nutrients are taken in not only through the roots, but also through leaf surfaces, so they are more immediately effective, resulting in a higher growth rate.

The temperature is increased and the level of lighting raised with artificial lighting; carbon dioxide is sprayed and oxygen pumped in, making plant growth several times faster than in field cultivation.

However, any product grown in such an artificial environment is a far cry from product grown under natural conditions. True, freshly colored melons with a beautifully networked skin and a sweet taste and fragrance can be produced, as can large red tomatoes and supple green cucumbers of good texture. But it is a mistake to think of these as good for man. Grown unnaturally as they are, these products are inferior in quality, although perhaps in ways unknown to man. Nature has struck back fiercely against this affront by technology, in the form of increased insect damage. Predictably, the response by man has been an agriculture increasingly dependent on pesticides and fertilizers.

Artificial cultivation leads ultimately to the total synthesis of food. The creation of factories for purely chemical food synthesis that will render farms and gardens unnecessary is already underway. This will make of agriculture an activity entirely unrelated to nature.

The synthesis of urea has enabled man to produce any organic material he wishes. Protein synthesis enables man-made meat to be fabricated from various materials. Butter and cheese can be made from petroleum. Sooner or later, as further progress is made in research on photosynthesis, man will surely learn how to synthesize starch. He may even succeed one day in doing this by the saccharification of wood and oil.

Man has learned how to synthesize nucleic acid and cellular proteins and nuclei, and is beginning to synthesize and recombine genes and chromosomes. He has even begun thinking that he can control life itself. Not only that. As the notion has settled in that he may soon be able to alter all living things in any way he pleases, man has begun fancying himself as the Creator. Yet all that he learns, all that he performs and creates with science, is a mere imitation of nature and propels him further along the path to suicidal self-destruction.

Production Costs Are Not Coming Down

It is a mistake to believe that progress in agricultural technology will lower production costs and make food less expensive. Suppose that some entrepreneur decided to grow rice and vegetables in a large building right at the center of a major city. He would make full spatial use of the building in three dimensions, fully equipping it with central heating and air conditioning, artificial lighting, and automatic spraying devices for carbon dioxide and nutrient solutions.

Now, would such systematized agriculture involving automated production under the watchful eyes of a single technician really provide people with fresh, inexpensive, and nutritious vegetables? A vegetable factory like this cannot be built and run without considerable outlays for capital and materials, so it is only natural to expect the vegetables thus produced to be expensive. However efficient and modern it may be, such a plant cannot possibly grow produce more cheaply than crops grown naturally with sunlight and soil.

Nature produces without calling for supplies or remuneration, but human effort always demands payment in return. The more sophisticated the equipment and facilities, the higher the costs. And man never knows where to stop. When a highly efficient robot is developed, people applaud, saying that efficient production is here at last. But their joy is short-lived, for soon they are dissatisfied again and demanding even more advanced and efficient technology. Everyone seems intent on lowering production costs, but these costs skyrocketed before anyone knew what was happening.

Equally mistaken is the notion that food can be produced cheaply and in large quantity with microorganisms such as chlorella and yeast. Science cannot produce something from nothing. Invariably, the result is a decrease in production rather than an increase, giving a high-cost product.

People brought up eating unnatural food develop into artificial, anti-natural human beings with an unnatural body prone to disease and an unnatural way of thinking. There exists the frightful possibility that the transfiguration of agriculture may result in the perversion of far more than just agriculture.

Increased Production Has Not Brought Increased Yields

When talk everywhere turned to increasing food production, most people believed that raising yields and productivity through scientific techniques would enable man to produce larger, better, more plentiful food crops. Yet, larger harvests have not brought greater profits for farmers. In many cases, they have even resulted in losses.

Most high-yield farming technology in use today does not increase net profits. At fault are the very practices thought to be vital to increasing yields: the heavy application of chemical fertilizers and pesticides, and indiscriminate mechanization. But although these may be useful in reducing crop losses, they are not effective techniques for increasing productivity. In fact, such practices hurt productivity. They appear to work because:

1) Chemical fertilizers are effective only when the soil is dead.
2) Pesticides are effective only for protecting unhealthy plants.
3) Farm machinery is useful only when one has to cultivate a large area.

Another way of saying the same thing is that these methods are ineffective or even detrimental on fertile soil, healthy crops, and small fields. Chemical fertilizers can increase yields when the soil is poor to begin with and produces only 4 to 5 bushels of rice per quarter-acre. Even then, heavy fertilization produces an average rise in yield of not more than about 2 bushels over the long term. Chemical fertilizers are truly effective only on soil abused and wasted through slash-and-burn agriculture.

Adding chemical fertilizer to soil that regularly produces 7 to 8 bushels of rice per quarter-acre has very little effect, while addition to fields that yield 10 bushels may even hurt productivity. Chemical fertilizer is thus of benefit only as a means for preventing a decline in yields. Green manure—nature's own fertilizer—and animal manure were cheaper and safer methods of increasing yields.

The same is true of pesticides. What sense can there be in producing unhealthy rice plants and applying powerful pesticides anywhere up to ten times a year? Before investigating how well pesticides kill harmful insects and how well they prevent crop losses, scientists should have studied how the natural ecosystem is destroyed by these pesticides and why crop plants have weakened. They should have investigated the causes underlying the disruption in the harmony of nature and the outbreak of pests, and on the basis of these findings decided whether pesticides are really needed or not.

By flooding the paddy fields and breaking up the soil with tillers until it hardens to the consistency of adobe, rice farmers have created conditions that make it impossible to raise crops without tilling, and in the process have deluded themselves into thinking this to be an effective and necessary part of farming. Fertilizers, pesticides, and farm machinery all appear convenient and useful in raising productivity. However, when viewed from a broader perspective, these kill the soil and crops, and destroy the natural productivity of the earth.

"But after all," we are often told, "along with its advantages, science also has its disadvantages." Indeed, the two are inseparable; we cannot have one without the other. Science can produce no good without evil. It is effective only at the price of the destruction of nature. This is why, after man has maimed and disfigured nature, science appears to give such striking results—when all it is doing is repairing the most extreme damage.

Productivity of the land can be improved through scientific farming methods only when its natural productivity is in decline. These are regarded as high-yielding practices only because they are useful in stemming crop losses. To make matters worse, man's efforts to return conditions to their natural state are always incomplete and accompanied by great waste. This explains the basic energy extravagance of science and technology.

Nature comes into being entirely of itself. In its eternal cycles of change, never is there the slightest extravagance or waste. All the products of the human intellect—which has strayed far from the bosom of nature—and all man's labors are doomed to end in vain.

Before rejoicing over the progress of science, we should lament those conditions that have driven us to depend on its helping hand. The root cause for the decline of the farmer and crop productivity lie with the development of scientific agriculture.

Energy-Wasteful Modern Agriculture

The claim is often made that scientific agriculture has a high productivity, but if we calculate the energy efficiency of production, we find that this decreases with mechanization. Table 1.1 compares the amount of energy expended directly in rice production using five different methods of farming: natural farming, farming with the help of animals, and lightly, moderately, and heavily mechanized agriculture. Natural farming requires only one man-day of labor to recover 130 pounds of rice, or 200,000 kilocalories of food energy, from a quarter-acre of land. The energy input needed to recover 200,000 kilocalories from the land in this way is the 2,000 kilocalories required to feed one farmer for one day. Cultivation with horses or oxen requires an energy input five to ten times as great, and mechanized agriculture calls for an input of from ten to fifty times as much energy. Since the efficiency of rice production is inversely proportional to the energy input, scientific agriculture requires an energy expenditure per unit of food produced up to fifty times that of natural farming.

The youths living in the mud-walled huts of my citrus orchard have shown me that a person's minimum daily calorie requirement is somewhere about 1,000 calories for a "hermit's diet" of brown rice and sesame-salt, and 1,500 calories on a diet of brown rice and vegetables. This is enough to do a farmer's work—about one-tenth of a horsepower.

Table 1.1 Direct energy input in rice production, given as number of kilocalories required to produce 1,300 pounds (22 bushels) of rice on a quarter-acre.

	Natural farming	Farming with animals (ca. 1950)[1]	Small-scale mechanized agriculture (ca. 1960)	Medium-scale mechanized agriculture (ca. 1970)	Large-scale mechanized agriculture (ca. 1980)[2]	Remarks
Human labor	10–20	25	20	12	—	kilocalories in diet
Animal labor	0	6	4	0	0	
Machinery	hand tools	22	80	350	—	
Fertilizer	0	40	75	54	—	
Pesticides	0	11	25	72	—	kcal. of rice energy
Fuel	0	2	10	45	—	
Total	10–20	96	214	533	1,000	
Energy input*	0.1–0.2	1	2	5	10	Assuming 200,000 kcal. per 1300 lbs. of rice
Energy output** / Energy input	100–200	20	10	4	2	

*Energy input for farming with animals=1
**Ratio of energy from harvested rice to energy input
1) Dates apply to Japan   2) Estimate

At one time, people believed that using horses and oxen would lighten the labor of men. But contrary to expectations, our reliance on these large animals has been to our disadvantage. Farmers would have been better off using pigs and goats to plow and turn the soil. In fact, what they should have done was to leave the soil to be worked by small animals—chickens, rabbits, mice, moles, and even worms. Large animals only appear to be useful when one is in a hurry to get the work done. We tend to forget that it takes over two acres of pasture to feed just one horse or cow. This much land could feed fifty or even a hundred people if one made full use of the powers of nature. Raising livestock has clearly taken its toll on man. The reason India's farmers are so poor today is that they raised large numbers of cows and elephants which ate up all the grass, and dried and burned the droppings as fuel. Such practices have depleted soil fertility and reduced the productivity of the land.

Livestock farming today is of the same school of idiocy as the fish-farming of yellowtails. Raising one yellowtail to a marketable size requires ten times its weight in sardines. Similarly, a silver fox consumes ten times its weight in rabbit meat, and a rabbit ten times its weight in grass. What an incredible waste of energy to produce a single silver fox pelt! People have to work ten times as hard to eat beef grain, and they had better be prepared to work five times as hard if they want to nourish themselves on milk and eggs.

Farming with the labor of animals therefore helps satisfy certain cravings and desires, but increases man's labor many times over. Although this form of agriculture appears to benefit man, it actually places him at the service of his livestock. In raising cattle or elephants as members of the farming household, the peasants of Japan and India impoverished themselves to provide their livestock with the calories they needed.

Mechanized farming is even worse. Instead of reducing the farmer's work, mechanization enslaves him to his equipment. To the farmer, machinery is the largest domestic animal of all—a great guzzler of oil, a consumer good rather than a capital good. At first glance, mechanized agriculture appears to increase the productivity per worker and thus raise income. However, quite to the contrary, a look at the efficiency of land utilization and energy consumption reveals this to be an extremely destructive method of farming.

Man reasons by comparison. Thus he thinks it better to have a horse do the plowing than a man, and thinks it more convenient to own a ten-horsepower tractor than to keep ten horses—why, if it costs less than a horse, a one-horse-power motor is a bargain! Such thinking has accelerated the spread of mechanization and appears reasonable in the context of our currency-based economic system. But the progressively inorganic character and lowered productivity of the land resulting from mass-production type farming operations, the economic disruption caused by the excessive input of energy, and the increased sense of alienation deriving from such a direct antithesis to nature amounts to the blatant rejection of agriculture, however much this is called "progress."

Has mechanization really increased productivity and made things easier for the farmer? Let us consider the changes mechanization has brought about in tilling practices.

A two-acre farmer who purchases a 30-horsepower tractor will not magically become a 50-acre farmer unless the amount of land in his care increases. If the land under cultivation is limited, mechanization only lowers the number of laborers required. This surplus manpower begets leisure. Applying this excess energy to some other work increases income, or so the reasoning goes. The problem, however, is that this extra income cannot come from the land. In fact, the yield from the land will probably decrease while the energy requirements skyrocket. In the end, the farmer is driven from his fields by his machinery. The use of machinery may make working the fields easier, but revenue from crop production has shrunk. Yet taxes are not about to decrease, and the costs of mechanization continue to climb by leaps and bounds. This is where things stand for the farmer.

The reduction in labor brought about by scientific farming has succeeded only in forcing farmers off the land. Perhaps the politician and consumer think the ability of a smaller number of workers to carry out agricultural production for the nation is indicative of progress. To the farmer, however, this is a tragedy, a preposterous mistake. For every tractor operator, how many dozens of farmers are driven off the land and forced to work in factories making agricultural implements and fertilizer—which would not be needed in the first place if natural farming were used.

Machinery, chemical fertilizers, and pesticides have drawn the farmer away from nature. Although these useless products of human manufacture do not raise the yields of his land, because they are promoted as tools for making profits and boosting yields, he labors under the illusion that he needs them. What is more, their use has wrought great destruction on nature, robbing it of its powers and leaving man no choice but to tend vast fields by his own hand. This in turn has made large machinery, high-grade compound fertilizers, and powerful poisons indispensable. And the same vicious cycle goes on and on without end.

The farmer has not found stability with the ever-increasing scale of agricultural operations. Farms in Europe are ten times larger, and in the United States one hundred times larger, than the 6- to 7- acre farms common to Japan. Yet farmers in Europe and the U.S. are, if anything, even more insecure than Japanese farmers. It is only natural that farmers in the West who question the trend toward large-scale mechanized agriculture have sought an alternative in Eastern methods of organic farming. However, as they have come to realize also that traditional agriculture with farm animals is not the road to salvation, these farmers have begun searching frantically for the road leading toward natural farming.

Laying to Waste the Land and Sea

The modern livestock and fishing industries are also basically flawed. Everyone unquestioningly assumed that, by raising poultry and livestock and by fish-farming, our diet would improve, but no one had the slightest suspicion that the production of meat would ruin the land and the production of fish would pollute the seas.

In terms of caloric production and consumption, someone will have to work at least twice as hard if he wants to eat eggs and milk rather than grains and vege-tables. If he likes meat, he will have to put out seven times the effort. Because it is so energy-inefficient, modern livestock farming cannot be considered as "produc-tion" in a basic sense. In fact, true production efficiency has become so low and man has been driven to such extremes of toil and labor that he is even attempting to increase the efficiency of livestock production by raising large, genetically improved breeds.

The Japanese Bantam is a breed of chicken native to Japan. Leave it to roam about freely and it lays just one small egg every other day—low productivity by most standards. But although this chicken is not an outstanding egg-layer, it is in fact very productive. Take a breeding pair of Bantams, let them nest every so often, and before you know it they will hatch a clutch of chicks. Why, within a year's time, your original pair of chickens will have grown to ten or twenty chickens that together will lay many times more eggs each day than the best variety of White Leghorn. The Bantams are very efficient calorie producers be-cause they feed themselves and lay eggs on their own, literally producing some-thing from nothing. Moreover, as long as the number of chickens remains appropriate for the space available, raising chickens in this way does not harm the land.

Genetically-upgraded White Leghorns raised in cages lay one large egg a day. Because they produce so many eggs, it is commonly thought that raising these in large numbers will provide people with lots of eggs to eat and also generate droppings that can be used to enrich the land. But in order for the chickens to lay so many eggs, they have to be given feed grain having twice the caloric value of the eggs produced. Such artificial methods of raising chickens are thus basically counterproductive; instead of increasing calories, they actually cut the number of calories in half. Restoration of the wastes to the land is not easy, and even then, soil fertility is depleted to the extent of the caloric loss.

This is true not only for chickens but for pigs and cattle as well, where the efficiency is even worse. The ratio of energy output to input is 50% for broilers, 20% for pork, 15% for milk, and 8% for beef. Raising beef cattle cuts the food energy recoverable from land tenfold; people who eat beef consume ten times as much energy as people on a diet of rice. Few are aware of how our livestock industry, which raises cattle in indoor stalls with feed grain shipped from the United States, has helped deplete American soil. Not only are such practices uneconomical, they amount essentially to a campaign to destroy vegetation on a global scale.

Nonetheless, people persist in believing that raising large numbers of chickens that are good egg layers or improved breeds of pigs and cattle with a high feed conversion efficiency in enclosures is the only workable approach to mass-production and is intelligent, economical livestock farming. The very opposite is true. Artifical livestock practices consisting essentially of the conversion of feed into eggs, milk, or meat are actually very energy-wasteful. In fact, the larger and more highly improved the breed of animal being raised, the greater the energy input required and the greater the effort and pains that must be taken by the farmer.

The question we must answer then is, What should be raised, and where? First we must select breeds that can be left to graze the mountain pastures. Raising large numbers of genetically improved Holstein cows and beef cattle in indoor pens or small enclosures on concentrated feed is a highly risky business for both man and livestock alike. Moreover, such methods yield higher rates of energy loss than other forms of animal husbandry. Native breeds and varieties such as Jersey cattle, which are thought to be of lower productivity, actually have a higher feed efficiency and do not lead to depletion of the land. Being closer to nature, the wild boar and the black Berkshire pig are in fact more economical than the supposedly superior white Yorkshire breed. Profits aside, it would be better to raise small goats than dairy cattle. And raising deer, boars, rabbits, chickens, wildfowl, and even edible rodents, would be even more economical—and better protect nature—than goats.

In a small country like Japan, rather than raising large dairy cattle, which merely impoverishes the soil, it would be far wiser for each family to keep a goat. Breeds that are better milk producers but basically weak, such as Saanen, should be avoided, and strong native varieties that can live on roughage raised. The goat is called the poor man's cattle because it takes care of itself and also provides milk, but it is in fact inexpensive to raise and does not weaken the productivity of the land.

If poultry and livestock are to truly benefit man, they must be capable of feeding and fending for themselves under the open sky. Only then will food become naturally plentiful and contribute to man's well-being.

In my idealized picture of livestock farming, I see bees busily making the rounds of clover and vegetable blossoms thickly flowering beneath trees laden heavy with fruit; I see semi-wild chickens and rabbits frolicking with dogs in fields of growing wheat, and great numbers of ducks and mallards playing in the rice paddy; at the foot of the hills and in the valleys, black pigs and boars grow fat on worms and crayfish, and from time to time goats peer out from the thickets and trees.

This scene might be taken from an out-of-the-way hamlet in a country untarnished by modern civilization. The real question for us is whether to view it as a picture of primitive, economically disadvantaged life or as an organic partnership between man, animal, and nature. An environment comfortable for small animals is also an ideal setting for man.

It takes 200 square yards of land to support one human being living on grains, 600 square yards to support someone living on potatoes, 1,500 square yards for someone living on milk, 4,000 square yards for someone living on pork, and 10,000 square yards for someone subsisting entirely on beef. If the entire human population on earth were dependent on a diet of just beef, humanity would have already reached its limits of growth. The world population could grow to three times its present level on a diet of pork, eight times on a milk diet, and twenty times on a potato diet. On a diet of just grains, the earth's carrying capacity is sixty times the current world population.

One need look only at the United States and Europe for clear evidence that beef impoverishes the soil and denudes the earth.

Modern fishing practices are just as destructive. We have polluted and killed the seas that were once fertile fishing grounds. Today's fishing industry raises expensive fish by feeding them several times their weight in smaller fish while rejoicing at how abundant fish have become. Scientists are interested only in learning how to make bigger catches or increasing the number of fish, but viewed in a larger context, such an approach merely speeds the decline in catches. Protecting seas in which fish can still be caught by hand should be a clear priority over the development of superior methods for catching fish. Research on breeding technology for shrimp, sea bream, and eels will not increase the numbers of fish. Such misguided thinking and efforts are not only undermining the modern agricultural and fishing industries, they will also someday spell doom for the oceans of the world.

As with modern livestock practices that run counter to nature, man has tricked himself into believing that he can improve the fishing industy through the development of more advanced fish farming methods while at the same time perfecting fishing practices that destroy natural reproduction. Frankly, I am frightened at the dangers posed by treating fish with large doses of chemicals to prevent pelagic diseases that break out in the Inland Sea as a result of pollution caused by the large quantities of feed strewn over the water at the many fish farming centers on the Sea. It was no laughing matter when a rise in demand for sardines as feed for yellowtails resulted in a curious development recently: an acute shortage of sardines that made the smaller fish a luxury item for a short while.

Man ought to know that nature is fragile and easily harmed. It is far more difficult to protect than everyone seems to think, and once it has been destroyed, nature cannot be restored.

The way to enrich man's diet is easy. It does not entail mass growing or gathering, but it does require man to relinquish human knowledge and action and allow nature to restore its natural bounty. Indeed, there is no other way.

The Illusions
of Natural Science | 2

1. The Errors of the Human Intellect

Scientific agriculture developed early in the West as one branch of the natural sciences, which arose in Western learning as the study of matter. The natural sciences took a materialistic viewpoint that interpreted nature analytically and dialectically.

This was a consequence of Western man's belief in a man-nature dichotomy. In contrast to the Eastern view that man should seek to become one with nature, Western man used discriminating knowledge to place man in opposition to nature and attempted, from that vantage point, a detached interpretation of nature. For he was convinced that the human intellect can cast off subjectivity and comprehend nature objectively.

Western man firmly believed nature to be an entity with an objective reality independent of human consciousness, an entity that man can know through observation, reductive analysis, and reconstruction. From these processes of destruction and reconstruction arose the natural sciences.

The natural sciences have advanced at breakneck speed, flinging us into the space age. Today, man appears capable of knowing everything about the universe. He grows ever more confident that, sooner or later, he will understand even phenomena as yet unknown. But what exactly does it mean for man to "know"? He may laugh at the folly of the proverbial frog in the well, but is unable to laugh off his own ignorance before the vastness of the universe. Although man, who occupies but one small corner of the universe, can never hope to fully understand the world in which he lives, he persists nonetheless in the illusion that he has the cosmos in the palm of his hand. Man is not in a position to know nature.

Nature Must Not Be Dissected

Scientific farming first arose when man, observing crops as they grew, came to know the crops and later became convinced that he could grow them himself. Yet has man really known nature? Has he really grown crops and lived by the fruit of his own labor? Man looks at a stalk of wheat and says he knows what that wheat is. But does he really know wheat and is he really capable of growing it? Let us examine the process by which man thinks he can know things.

Man believes that he has to fly off into outer space to learn about space, or that he must travel to the moon to know the moon. In the same way, he thinks that to know a stalk of wheat, he must first take it in his hand, dissect it, and analyze it. He thinks that the best way to learn about something is to collect and assemble as much data on it as possible. In his efforts to learn about nature, man has cut it up into little pieces. He has certainly learned many things in this way, but what he has examined has not been nature itself.

Man's curiosity has led him to ask why and how the winds blow and the rain falls. He has carefully studied the tides of the sea, the nature of lightning, and the plants and animals that inhabit the fields and mountains. He has extended his inquiring gaze into the tiny world of microorganisms, into the realm of minerals and inorganic matter. Even the sub-microscopic universe of molecules and atomic and subatomic particles have come under his scrutiny. Detailed research has pressed forth on the morphology, physiology, ecology, and every other conceivable aspect of a single flower, a single stalk of wheat.

Even a single leaf presents infinite opportunities for study. The collection of cells that together form the leaf; the nucleus of one of these cells, which harbors the mystery of life; the chromosomes that hold the key to heredity; the question of how chlorophyll synthesizes starch from sunlight and carbon dioxide; the unseen activity of roots at work; the uptake of various nutrients by the plant; how water rises to the tops of tall trees; the relationships with various components and microorganisms in the soil; how these interact and change when absorbed by the roots and what functions they serve—these are but a few of the inexhaustible array of topics scientific research has pursued.

But nature is a living, organic whole that cannot be divided and subdivided. When it is separated into two complementary halves and these divided again into four, when research becomes fragmented and specialized, the unity of nature is lost.

The diagram in Figure 2.1 is an attempt to illustrate the interplay of factors, or elements, that determine yields in rice cultivation.

Originally, the elements determining yield were not divided and separate. All were joined in perfect order under a single conductor's baton and resonated together in exquisite harmony. Yet, when science inserted its scalpel, a complex and horrendously chaotic array of elements appeared. Science has done nothing other than to peel the skin off a beautiful woman and reveal a bloody mass of tissue. What a miserable, wasted effort.

Nowadays, plants can be made to bloom in all seasons. Stores display fruits and vegetables throughout the year, so that one almost forgets whether it is summer or winter anymore. This is the result of chemical controls that have been developed to regulate the time of bud formation and differentiation.

Confident of his ability to synthesize the proteins that make up cells, man has even challenged the "ultimate" secret—the mystery of life itself. Whether he will succeed in synthesizing cells depends on his ability to synthesize nucleic acids, this being the last major hurdle to the synthesis of living matter. The synthesis of simple forms of life is now just a matter of time. This was first anticipated when the notion of a fundamental difference between living and non-living matter was laid to rest with the discovery of bacteriophages, the confirmation—in subsequent research on viral pathogens—of the existence of non-living matter that multiplies, and the first attempts to synthesize such matter.

Following his interests blindly, man is intently at work on the synthesis of life without knowing what the successful creation of living cells means or the repercussions it might have. Nor is this all. Carried along by their own momentum, scientists have even begun venturing into chromosome synthesis. Soon after the disclosure that man had synthesized life came the announcement that the synthesis and modification of chromosomes has become possible through genetic recombina-

Fig. 2.1   The factors of rice cultivation.

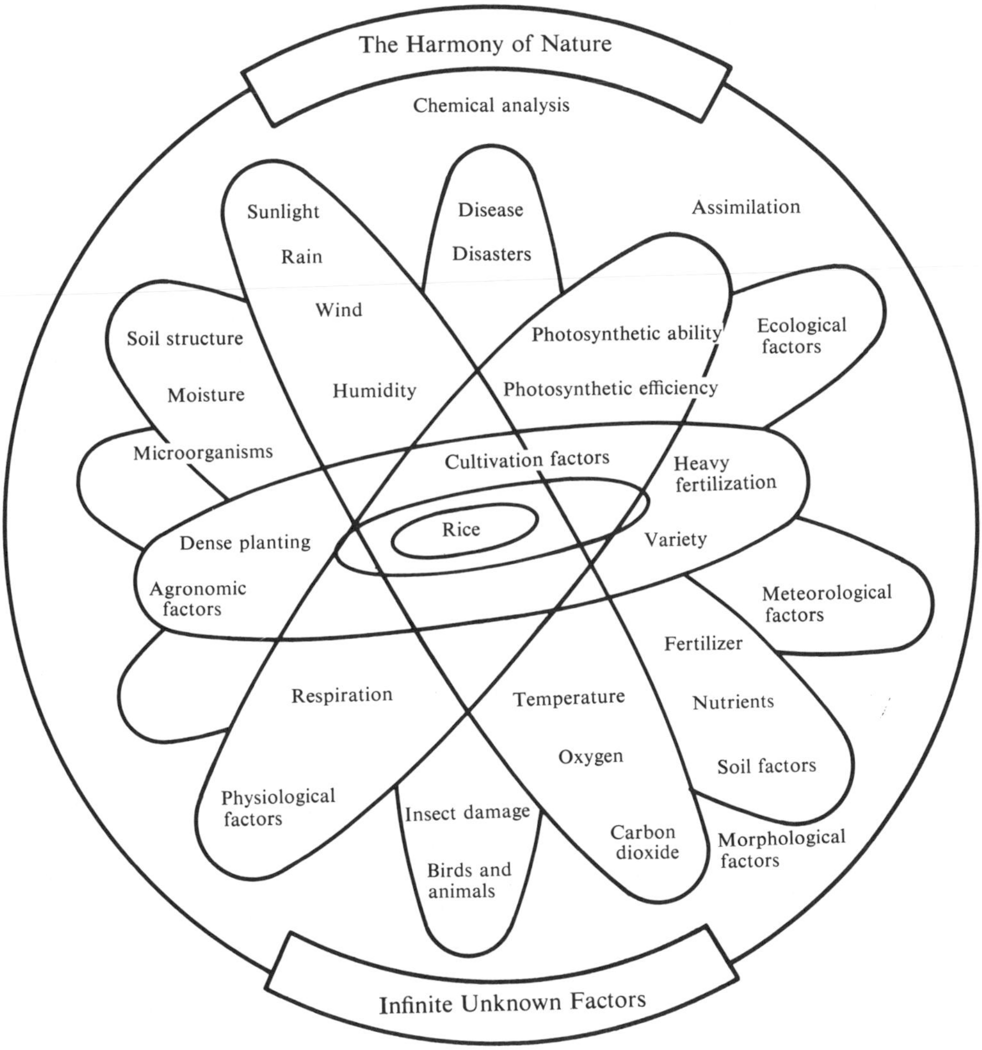

The Harmony of Nature

Chemical analysis

Sunlight

Disease

Assimilation

Rain

Disasters

Wind

Photosynthetic ability

Ecological factors

Soil structure

Moisture

Humidity

Photosynthetic efficiency

Microorganisms

Cultivation factors

Heavy fertilization

Dense planting

Rice

Variety

Agronomic factors

Meteorological factors

Fertilizer

Respiration

Temperature

Nutrients

Oxygen

Soil factors

Physiological factors

Insect damage

Carbon dioxide

Morphological factors

Birds and animals

Infinite Unknown Factors

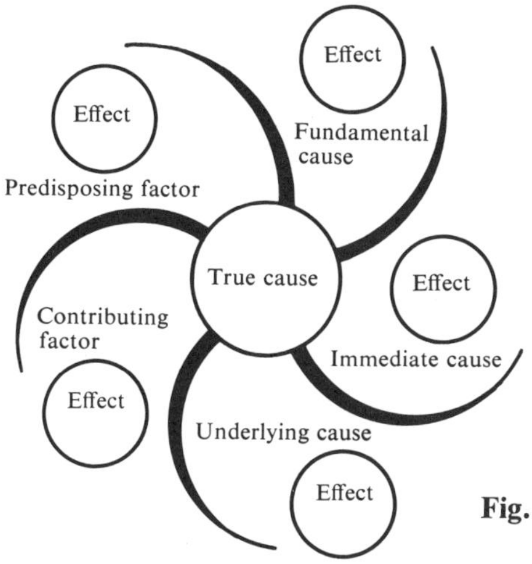

Effect

Effect

Fundamental cause

Predisposing factor

True cause

Effect

Contributing factor

Immediate cause

Effect

Underlying cause

Effect

Fig. 2.2   Relationship between cause and effect.

tion. Man can already create and alter living organisms like the Creator. We are about to enter an age in which scientists will create organisms that have never before appeared on the face of the earth. Following test-tube babies, we will see the creation of artificial human beings, monsters, and enormous crops. In fact, these have already begun to appear.

Granted, one certainly does get the impression that great advances have been made in human understanding, that man has come to know all things in nature and, by using and adapting such knowledge, has accelerated progress in human life. Yet, there is a catch to all of this. Man's awareness is intrinsically imperfect, and this gives rise to errors in human understanding.

When man says that he is capable of knowing nature, to "know" does not mean to grasp and understand the true essence of nature. It means only that man knows that nature which he is able to know.

Just as the world known to a frog in a well is not the entire world but only the world within that well, so the nature that man can perceive and know is only that nature which he has been able to grasp with his own hands and his own subjectivity. But of course, this is not true nature.

The Maze of Relative Subjectivity

When people want to know what Okuninushi no Mikoto, the Shinto god of healing, carries around in the huge sack on his shoulder, they immediately open the sack and thrust their hands in. They think that to understand the interior of the sack, they must know its contents. Supposing they found the sack to be filled with all sorts of strange objects made of wood and bamboo. At this point, most people would begin to make various pronouncements: "Why this no doubt is a tool used by travelers." "No, it is a decorative carving." "No, it is most definitely a weapon." And so forth. Yet the truth, known only to Okuninushi no Mikoto, is that the object is an instrument fashioned by him for his amusement. And moreover, because it is broken, he is carrying it around in his sack merely for use as kindling.

Man jumps into that great sack called nature, and grabbing whatever he can, turns it over and examines it, asking himself what it is and how it works, and drawing his own conclusions about what purpose nature serves. But no matter how careful his observations and reasoning, each and every interpretation carries the risk of causing grievous error because man cannot know nature any more than he can know the uses for the objects in Okuninushi no Mikoto's sack.

Yet man is not easily discouraged. He believes that, even if it amounts to the same absurdity as jumping into Okuninushi no Mikoto's sack and guessing at the objects inside, man's knowledge will broaden without limit; simple observations will start the wheels of reason and inference turning.

For example, man may see some shells attached to a piece of bamboo and mistake it for a weapon. When further investigation reveals that rapping the shells against the bamboo makes an interesting sound, he will conclude this to be a musical instrument, and will infer from the curvature of the bamboo that it must be worn dangling from the waist while dancing. With each step in this line of reasoning, he will believe himself that much closer to the truth.

Just as he believes that he can know Okuninushi no Mikoto's mind by studying the contents of his sack, so man believes that, by observing nature, he can learn the story of its creation, and can in turn become privy to its very designs and purpose. But this is a hopeless illusion, for man can know the world only by stepping outside of the sack and meeting face-to-face with the owner.

A flea born and raised in the sack without ever having seen the world outside will never be able to guess that the object in the sack is an instrument that is hung from Okuninushi no Mikoto's belt, no matter how much it studies the object. Similarly, man, who is born within nature and will never be able to step outside of the natural world, can never understand all of nature merely by examining that part of nature around him.

Man's answer to this is that, although he may not be able to view the world from without, if he has the knowledge and ability to explore the furthest reaches of the vast, seemingly boundless universe and is able at least to learn what there is and what has happened in this universe, is not this enough? Has not man learned, sooner or later, everything that he wished to learn? That which is unknown today will become known tomorrow. This being the case, there is nothing man cannot know.

Even if he were to spend his entire life within a sack, provided he were able to learn everything about the inside of the sack, would this not be enough? Is not the frog in the well able to live there in peace and tranquility? What need has it for the world outside the well?

Man watches nature unfold about him; he examines it and puts it to practical use. If he gets the expected results, he has no reason to call into question his knowledge or actions. There being nothing to suggest that he is in error, does not this mean that he has grasped the real truth about the world?

He assumes an air of nonchalance: "I don't know what lies outside the world of the unknown; maybe nothing. This goes beyond the sphere of the intellect. We'd be better off leaving inquiries into a world that may or may not exist to those men of religion who dream of God."

But who is it that is dreaming? Who is it that is seeing illusions? And knowing the answer to this, can we enjoy true peace of mind? No matter how deep his understanding of the universe, it is man's subjectivity that holds up the stage on which his knowledge performs. But just what if his subjective view were all wrong?

Before laughing at blind faith in God, man should take note of his blind faith in himself.

When man observes and judges, there is only the thing called "man" and the thing being observed. It is this thing called "man" that verifies and believes in the reality of an object, and it is man who verifies and believes in the existence of this thing called "man." Everything in this world derives from man and he draws all the conclusions. In which case, he need not worry about being God's puppet. But he does run the risk of acting out a drunken role on the stage supported by the crazed subjectivity of his own despotic existence.

"Yes," persists the scientist, "man observes and makes judgments, so one cannot deny that subjectivity may be at work here. Yet his ability to reason enables man to divest himself of subjectivity and see things objectively as well. Through repeated inductive experimentation and reasoning, man has resolved all things into patterns of association and interaction. The proof that this was no mistake lies about us, in the airplanes, automobiles, and all the other trappings of modern civilization."

But if, on taking a better look at this modern civilization of ours, we find it to be insane, we must conclude that the human intellect which engendered it is also insane. It is the perversity of human subjectivity that gave rise to our ailing modern age. Indeed, whether one views the modern world as insane or not may even be a criterion of one's own sanity. We have already seen, in Chapter 1, how perverted agriculture has grown.

Are airplanes really fast, and cars truly a comfortable way to travel? Isn't our magnificent civilization nothing more than a toy, an amusement? Man is unable to see the truth because his eyes are veiled by subjectivity. He has looked at the green of trees without knowing true green, and has known the color crimson without seeing crimson itself. That has been the source of all his errors.

Non-Discriminating Knowledge

The statement that science arose from doubt and discontent is often used as an implied justification of scientific inquiry, but this in no way justifies it. On the contrary, when confronted with the havoc wrought by science and technology on nature, one cannot help feeling disquiet at this very process of scientific inquiry that man uses to separate and classify his doubts and discontents.

An infant sees things intuitively. When observed without intellectual discrimination, nature is entire and complete—a unity. In this non-discriminating view of creation, there is no cause for the slightest doubt or discontent. A baby is satisfied and enjoys peace of mind without having to do anything.

The adult mentally picks things apart and classifies them; he sees everything as imperfect and fraught with inconsistency. This is what is meant by grasping things dialectically. Armed with his doubts about "imperfect" nature and his discontent, man has set forth to improve upon nature and vainly calls the changes he has brought about "progress" and "development."

People believe that as a child grows into adulthood his understanding of nature deepens and through this process he becomes able to contribute to progress and development in this world. That this "progress" is nothing other than a march toward annihilation is clearly shown by the spiritual decay and environmental pollution that plague the developed nations of the world.

When a child living in the country comes across a muddy rice field, he jumps right in and plays in the mud. This is the simple, straightforward way of a child who knows the earth intuitively. But a child raised in the city lacks the courage to jump into the field. His mother has constantly been after him to wash the grime from his hands, telling him that dirt is filthy and full of germs. The child who "knows" about the "terrible germs" in the dirt sees the muddy rice field as unclean, an ugly and fearful place. Are the mother's knowledge and judgment really better than the unschooled intuition of the country child?

Hundreds of millions of microorganisms crowd each gram of soil. Bacteria are present in this soil, but so are other bacteria that kill these bacteria, and yet other bacteria that kill the killer bacteria. The soil contains bacteria harmful to man, but also many that are harmless or even beneficial to man. The soil in the fields under the sun is not only healthy and whole, it is absolutely essential to man. A child who rolls in the dirt grows up healthy. An unknowing child grows up strong.

What this means is that the knowledge that "there are germs in the soil" is more ignorant than ignorance itself. People would expect the most knowledgeable person on soil to be the soil scientist. But if, in spite of his extensive knowledge on soil as mineral matter in flasks and test tubes, his research does not allow him to know the joy of lying on the ground under the sun, he cannot be said to know anything about the soil. The soil that he knows is a discreet, isolated part of a whole. The only complete and whole soil is natural soil before it is broken down and analyzed, and it is the infant and child who best know, in their ingenuous way, what truly natural soil is.

The mother (science) who parades her partial knowledge implants in the child (modern man) a false image of nature. In Buddhism, knowledge that splits apart self and object and sets them up in opposition is called "discriminating knowledge," while knowledge that treats self and object as a unified whole is called "non-discriminating knowledge," the highest form of wisdom.

Clearly, the "discriminating adult" is inferior to the "non-discriminating child," for the adult only plunges himself into ever-deepening confusion.

2.   The Fallacies of Scientific Understanding

The Limits to Analytical Knowledge

The scientific method consists of four basic steps. The first step is to consciously focus one's attention on something and to observe and examine it mentally. The second is to use one's powers of discernment and reasoning to set up a hypothesis and formulate a theory based on these observations. The third step is to empirically uncover a single principle or law from common results gathered from analogous experiences and repeated experimentation. And finally, when the results of inductive experimentation have been applied and found to hold, the final step is to accept this knowledge as scientific truth and affirm its utility to mankind.

As this process begins with research that discriminates, breaks down, and analyzes, the truths it grasps can never be absolute and universal truths.

Thus scientific knowledge is by definition fragmented and incomplete; no matter how many bits of incomplete knowledge are collected together, they can never form a complete whole. Man believes that the continued dissection and deciphering of nature enable broad generalizations to be made which give a full picture of nature, but this only breaks nature down into smaller and smaller fragments and reduces it to ever greater imperfection.

The judgment by man that science understands nature and can use it to create a more perfect world has had the very opposite effect of making nature incomprehensible and has drawn man away from nature and its blessings, so that he now gladly harvests imitation crops far inferior to those of nature.

To illustrate, let us consider the scientist who brings a soil sample back to the laboratory for analysis. Finding the sample to consist of organic and inorganic matter, he divides the inorganic matter up into its components, such as nitrogen, potassium, phosphorus, calcium, and manganese, and studies, say, the pathways by which these elements are absorbed by plants as nutrients.

He then plants seeds in pots or small test plots to study how plants grow in this soil. He also carefully examines the relationships between microorganisms in the soil and inorganic soil components, and the roles and effects of these microorganisms.

The wheat that grows of its own accord from fallen seed on the open ground and the wheat planted and grown in laboratory pots are both identical, but man expends great time, effort, and resources to raise wheat, all because of the blind faith in his own ability to grow more and better wheat than nature. Why does he believe this?

Wheat growth varies with the conditions under which the wheat is grown. Noting a variation in the size of the heads of wheat, the scientist sets about to investigate the cause. He discovers that when there is too little calcium or magnesium in the soil within the pot, growth is poor and the leaves whither. When he artificially supplements the calcium or magnesium, he notes that growth speeds up and large grains form. Pleased with his success, the scientist calls his discovery scientific truth, and treats it as an infallible cultivation technique.

But the real question here is whether the lack of calcium or magnesium was a true deficiency. What is the basis for calling it a deficiency and is the remedy prescribed really in the best interests of man? When a field really is deficient in some component, the first thing done should be to determine the true cause of the deficiency. Yet science begins by treating the most obvious symptoms. If there is bleeding, it stops the bleeding. For a calcium deficiency, it immediately applies calcium.

If this does not solve the problem, then science looks further and any number of reasons may come to light: perhaps the over-application of potassium reduced calcium absorption by the plant or changed the calcium in the soil to a form that cannot be taken up by the plant.

This calls for a new approach. But behind every cause, there is a second and a third cause. Behind every phenomenon there is a main cause, a fundamental cause, an underlying cause, and contributing factors. Numerous causes and effects intermesh in a complex pattern from which the true cause cannot easily be extricated. Even so, man is confident of the ability of science to find the true cause through persistent and ever deeper investigation and to set up effective ways of coping with the problem. Yet, just how far can man go in his investigation of cause-and-effect?

There Is No Cause-and-Effect in Nature

Behind every cause lie countless other causes. Any attempt to trace these back to their sources only leads one further away from an understanding of the true cause.

When soil acidity becomes a problem, one jumps to the immediate conclusion that the soil does not contain enough lime. However this deficiency of lime may be due not to the soil itself, but to a more fundamental cause such as erosion of the soil resulting from repeated cultivation on ground exposed by weeding, or perhaps is related to the rainfall or temperature. And applying lime to treat soil acidity thought to result from insufficient lime may bring about excessive plant growth and increase acidity even further, in which case one ends up confusing cause with effect. Soil acidity control measures taken without understanding why the soil became acidic in the first place may be just as likely to prolong acidity as to reduce it.

Right after the war, I used large quantities of sawdust and wood chips in my orchard. Soil experts opposed this, saying that the organic acids produced when the wood rots would most likely make the soil acidic and that to neutralize it I would have to apply large quantities of lime. Yet the soil did not turn acid, so lime was not needed. What happens is that, when bacteria start decomposing the sawdust, organic acids are produced. But as the acidity rises, bacterial growth levels off and molds begin to flourish. When the soil is left to itself, the molds are eventually replaced by mushrooms and other fungi, which break the sawdust down to cellulose and lignin. The soil at this point is neither acidic nor basic, but hovers about a point of equilibrium.

The decision to counteract the acidity of rotting wood by applying lime only addresses the situation at a particular moment in time and under certain assumed conditions without a full understanding of the causal relationships involved. Nonintervention is the wisest course of action.

The same is true for crop diseases. Believing rice blast to be caused by the infiltration of rice blast bacteria, farmers are convinced beyond a doubt that the disease can be dispelled by spraying copper or mercury agents. However, the truth is not so simple. High temperatures and heavy rainfall may be contributing factors, as may the over-application of nitrogenous fertilizers. Perhaps flooding of the paddy during a period of high temperature weakened the roots, or the variety of rice being grown has a low resistance to rice blast disease.

Any number of interrelated factors may exist. Different measures may be adopted at different times and under different conditions, or a more comprehensive approach applied. But with a general acceptance of the scientific explanation for rice blast disease comes the belief that science is working on a way to combat the disease. Steady improvement in the pesticides used for the direct control of the disease has led to the present state of affairs where pesticides are applied several times a year as a sort of panacea.

But as research digs deeper and deeper, what was once accepted as plain and simple fact is no longer clear, and causes cease to be what they appear.

For instance, even if we know that excess nitrogenous fertilizer is a cause of rice blast disease, determining how the excess fertilizer relates to attack by rice blast

58

bacteria is no easy matter. If the plant receives plenty of sunlight, photosynthesis in the leaves speeds up, increasing the rate at which nitrogenous components taken up by the roots are assimilated as protein that nourishes the stem and leaves or is stored in the grain. But if cloudy weather ensues or the rice is planted too densely, individual plants may receive insufficient light or too little carbon dioxide, slowing photosynthesis. This may cause an excess of nitrogenous components to remain unassimilated in the leaves, making the plant susceptible to the disease.

Thus, an excess of nitrogenous fertilizer may or may not be the cause of rice blast disease. One can just as easily ascribe the cause to insufficient sunlight or carbon dioxide, or to the amount of starch in the leaves, but then it turns out that to understand how these factors relate to rice blast disease, we need to understand the process of photosynthesis. Yet modern science has not yet succeeded in fully unlocking the secrets of this process by which starch is synthesized from sunlight and carbon dioxide in the leaves of plants.

We know that rotting roots make a plant susceptible to rice blast, but the attempts of scientists to explain why this is so are less than convincing. This happens when the balance between the surface portion of the plant and its roots breaks down. Yet in trying to define what that balance is, we must answer why a weight inequilibrium in the roots as compared with the stalk and leaves makes the plant susceptible to attack by pathogens, what constitutes an "unhealthy" state, and other conundrums that ultimately leave us knowing nothing.

Sometimes the problem is blamed on a weak strain of rice, but again no one is able to define what "weak" means. Some scientists talk of the silica content and stalk hardness, while others define "weakness" in terms of physiology, genetics, or some other branch of scientific learning. In the end, we gradually fail to understand even those causes that appeared clear at first, and completely lose sight of the true cause.

When man sees a brown spot on a leaf, he calls it abnormal, and if he finds an unusual bacteria on that spot, he calls the plant diseased. His confident solution to rice blast disease is to kill the pathogen with pesticides. But in so doing he has not really solved the problem of blast disease. Without a grasp of the true cause of the disease, his solution cannot be a real solution. Behind each cause lies another cause, and behind that yet another. Thus what we view as a cause can also be seen as the result of another cause. Similarly, what we think of as an effect may become the cause of something else.

Fig. 2.3 **Effect may be traced back to cause, and cause to prior cause, in an endless chain of cause and effect.**

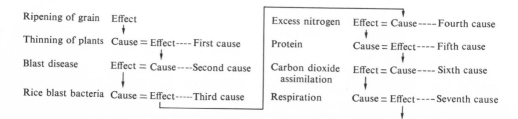

The rice plant itself may see blast disease as a protective mechanism that halts excessive plant growth and restores a balance between the surface and underground portions of the plant. The disease might even be regarded as a means by nature for preventing the overly dense growth of rice plants, thus aiding photosynthesis and assuring the full production of seed. In any case, rice blast disease is not the final effect, but merely one stage in the endless cycles of nature. It is both a cause as well as an effect.

Although cause and effect may be clearly discernible when observing an isolated event at a certain point in time, if one views nature from a broader spatial and temporal perspective, one sees a tangled confusion of causal relationships that defy unraveling into cause or effect. Even so, man thinks that by resolving this confusion down to its tiniest details and attempting to deal with these details at their most elementary level, he will be able to develop more precise and reliable solutions. But this scientific methodology and approach only results in the most circuitous and pointless efforts.

Viewed up close, organic causal relationships can be resolved into causes and effects, but when examined wholistically, no effects and causes are to be found. There is nothing to get a hold of, so all measures are futile. There is no cause-and-effect in nature. Nature has neither beginning nor end, before nor after, cause nor effect. Causality does not exist.

When there is no front or back, no beginning or end, but only what resembles a circle or a sphere, one could say that there is unity of cause and effect, but one could just as well say that cause and effect do not exist. This is my principle of non-causality.

To science, which examines this wheel of causality in parts and at close quarters, cause and effect exist. To the scientific mind trained to believe in causality, there most certainly is a way to combat rice blast bacteria. Yet when man, in his myopic way, perceives rice disease as a nuisance and takes the scientific approach of controlling the disease with a powerful bactericide, he proceeds from his first error that causality exists to subsequent errors. From his futile efforts he incurs further toil and misery.

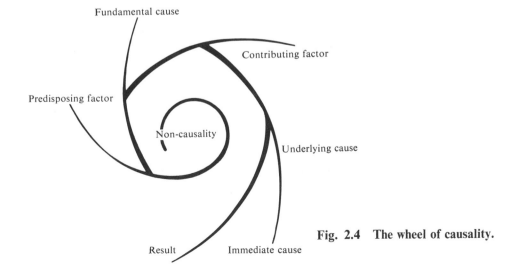

Fig. 2.4 The wheel of causality.

3. A Critique of the Laws of Agricultural Science

The Laws of Modern Agriculture

Certain generally accepted laws have been critical to the development of modern agricultural practices and serve as the foundation of scientific agriculture. These are the laws of diminishing returns, equilibrium, adaptation, compensation and cancellation, relativity, and the law of minimum. I would like to examine here the validity of each from the standpoint of natural farming. But before doing so, a brief description of these laws will help to show why each, when examined by itself, appears to stand up as an unassailable truth.

Law of diminishing returns: This law states, for example, that when one uses scientific technology to grow rice or wheat on a given plot of land and measures the resulting yields, the technology proves effective up to some upper limit, but exceeding this limit has the reverse effect of diminishing yields. Such a limit is not fixed in the real world, but changes with time and circumstances, so agricultural technology constantly seeks ways to break through it. Yet this law teaches that there are definite limits to returns, and that beyond a certain point additional effort is futile.

Equilibrium: Nature works constantly to strike a balance, to maintain an equilibrium. When this balance breaks down, forces come into effect that work to restore it. All phenomena in the natural world act to restore and maintain a state of equilibrium. Water flows from a high point to a low point, electricity from a high potential to a low potential. Flow ceases when the surface of the water is level, when there is no longer any difference in the electrical potential. The chemical transformation of a substance stops when chemical equilibrium has been restored. In the same way, all the phenomena associated with living organisms work tirelessly to maintain a state of equilibrium.

Adaptation: Animals live by adapting to their environment and crops similarly show the ability to adapt to changes in their growing conditions. Such adaptation is one type of activity aimed at restoring equilibrium in the natural world. The concepts of equilibrium and adaptation are thus intimately related and inseparable one from the other.

Compensation and cancellation: When rice is planted densely, the plants send out fewer tillers, and when it is planted sparsely, a larger number of stalks grow per plant. This is said to illustrate compensation. The notion of cancellation can be seen, for example, in the smaller heads of grain that result from increasing the number of stalks per plant, or in the smaller grains that form on heads of rice nourished to excessive size with heavy fertilization.

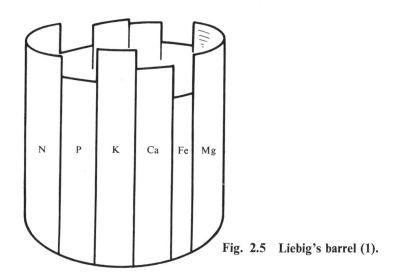

Relativity: Factors that determine crop yield are associated with other factors, and all change constantly in relation to each other. An interrelationship exists, for example, between the planting period and the quantity of seed sown, between the time and amount of fertilizer application, and between the number of seedlings and the spacing of plants. No particular amount of seed broadcast, quantity of fertilizer applied, or sowing period is decisive or critical under all conditions. Rather, the farmer constantly weighs one factor against another, making relative judgments that this variety of grain, that method of cultivation, or that type of fertilizer over there is right for such-and-such a period.

Law of minimum: This universally known law, first proposed by Justus von Liebig, a German chemist, may be said to have laid the foundation for the development of modern agriculture. It states that the yield of a crop is determined by the one constituent, of all those making up the yield, in shortest supply. Liebig illustrated this with a diagram now known as Liebig's barrel.

Fig. 2.5   Liebig's barrel (1).

The amount of water—or yield—the barrel holds is determined by that nutrient in shortest supply. No matter how large the supply of other nutrients, it is that nutrient of which there is the greatest scarcity that sets the upper limit on the yield.

A typical illustration of this principle would point out that the reason crops fail on volcanic soil in spite of the abundance of nitrogen, potassium, calcium, iron, and other nutrients is the scarcity of phosphates. Indeed, the addition of phosphate fertilizer often results in improved yields. In addition to tackling problems with soil nutrients, this concept has also been applied as a basic tool for achieving high crop yields.

Each of the above laws is treated and applied independently, yet are these really different and distinct from one another? My conclusion is that nature is an indivisible whole; all laws emanate from one source and return to Mu, or nothingness.

Scientists have examined nature from every conceivable angle, and have seen this unity as a thousand different forms. Although they recognize that these separate laws are intimately related and point in the same general direction, there is a world of difference between this realization and the awareness that all laws are one and the same.

One could read into the law of diminishing returns a force at work in nature that strives to maintain equilibrium by opposing and suppressing gradual increases in returns.

Compensation and cancellation are mutually antagonistic. The forces of cancellation act to negate the forces of compensation, by which mechanism nature seeks to maintain a balance.

Equilibrium and adaptability are, beyond any doubt, means of protecting the balance, order, and harmony of nature.

And if there is a law of the minimum, then there must also be a law of the maximum. In their search for equilibrium and harmony, plants have an aversion not only to nutrient deficiencies, but to deficiencies and excesses of anything.

Each one of these laws is nothing other than a manifestation of the great harmony and balance of nature. Each springs from a single source that draws them all together. What has misled man is that, when the same law emanates from a single source in different directions, he perceives each image as representing a different law.

Nature is an absolute void. Those who see nature as a point have gone one step astray, those who see it as a circle have gone two steps astray, and those who see breadth, matter, time, and cycles have wandered off into a world of illusion distant and divorced from true nature. The law of diminishing returns, which concerns gains and losses, does not reflect a true understanding of nature—a world without loss or gain.

When one has understood that there is no large or small in nature, only a great harmony, the notion of a minimum and a maximum nutrient also is reduced to a petty, circumstantial view. There was never any need for man to set into play his vision of relativity, to get all worked up over compensation and cancellation, or equilibrium and disequilibrium. Yet, agricultural scientists have drawn up elaborate hypotheses and added explanations for everything, leading farming further and further away from nature and upsetting the order and balance of the natural world.

Life on earth is a story of the birth and death of individual organisms, a cyclic history of the ascendance and fall, the thriving and failure, of communities. All matter behaves according to set principles—whether we are talking of the cosmic universe, the world of microorganisms, or the far smaller world of the atoms and electrons that make up nonliving and mineral matter. All things are in constant flux while preserving a fixed order; all things move in a recurrent cycle unified by some basic force emanating from a single source.

If we had to give this fundamental law a name, we could call it the "Dharmic Law That All Things Return to One." All things fuse into a circle, which reverts to a point, and the point to nothing. To man, it appears as if something has occurred and something has vanished, yet nothing is ever created or destroyed. This is not the same as the scientific law of the conservation of matter. Science maintains that destruction and conservation exist side by side, but ventures no further than this.

Fig. 2.6 All things return to one.

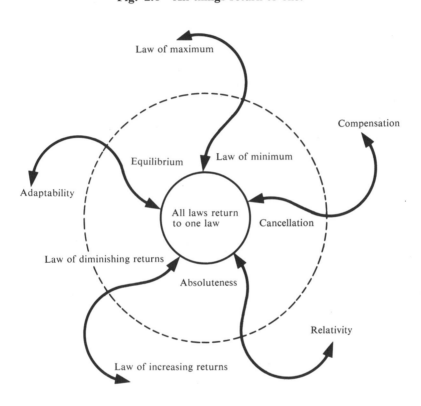

The different laws of agricultural science are merely scattered images, as seen through the prisms of time and circumstance, of this fundamental law that all things return to one. Because these laws all derive from the same source and were originally one, it is natural that they should fuse together in the same way that rice stalks join together at the base of the plant. Man might just as well have chosen to group together the law of diminishing returns, the law of minimum, and the law of compensation and cancellation and call these simply the "law of harmony." When we interpret this single law as several different laws, are we really explaining more of nature and achieving agricultural progress?

In his desire to know and understand nature, man applies numerous laws to it from many different perspectives. As would be expected, human knowledge deepens and expands, but man is sadly deceived in thinking that he draws closer to a true understanding of nature as he learns more about it. For he actually draws further and further away from nature with each new discovery and each fresh bit of knowledge.

These laws are fragments cut from the one law that flows at the source of nature. But this is not to say that if they were reassembled, they would form the original law. They would not.

Just as in the tale of the blind men and the elephant in which one blind man touches the elephant's trunk and believes it to be a snake and another touches one of the elephant's legs and calls it a tree, man believes himself capable of knowing the whole of nature by touching a part of it. There are limits to crop yields. There is balance and imbalance. Man observes the dualities of compensation and cancellation, of life and death, loss and gain. He notes nutrient excess and deficiency, abundance and scarcity, and from these observations derives various laws and pronounces them truths. He believes that he has succeeded in knowing and understanding nature and its laws, but what he has understood is nothing more than the elephant as seen by the blind men.

No matter how many fragmentary laws extracted from the single great law of nature are collected together, they can never add up to the source principle. That the nature observed through these laws differs fundamentally from true nature should come as no surprise. Scientific farming based on the application of such laws is vastly different from natural farming, which observes this basic principle of nature.

As long as natural farming stands on this unique law of nature, it is guaranteed truth and possesses eternal life. For although the laws of scientific farming may be useful in examining the status quo, they cannot be used to develop better cultivation techniques. These laws cannot boost rice yields beyond those attainable by present methods, and are useful only in preventing reduced yields.

When the farmer asks, "How many rice seedlings should I transplant per square yard of paddy," the scientist launches into a long-winded explanation of how the law of diminishing returns says that planting more than a certain number of seedlings does not increase yields, how compensation and cancellation are at work keeping seedling growth and the number of tillers within a given range to maintain an equilibrium, how too small a number of seedlings may be the limiting factor for yield and too large a number can also cause a decline in the harvested grain. At which point, the farmer asks with exasperation, "So what am I supposed to do?" Even the number of seedlings that should be planted varies with the conditions, and yet this has been the subject of endless research and debate.

No one knows how many stalks will grow from the seedlings planted in spring, or how this will affect yields in the fall. All one can do is theorize, after the harvest is in, that a smaller number of seedlings would have been better because of the high temperatures that summer, or that the combination of sparse planting and low temperatures were at fault for the low yields. These laws are of use only in explaining results, and cannot be of any help in reaching beyond what is currently possible.

In any discussion of increased production and high yields, the following are generally given as factors affecting yield:

Meteorological conditions	sunlight, temperature, humidity, wind strength, air, oxygen, carbon dioxide, hydrogen, etc.
Soil conditions	Physical: structure, moisture, air Chemical: inorganic, organic, nutrients, constituents
Biological conditions	animals, plants, microorganisms
Artificial conditions	Breeding Cultivation Manure and fertilizer application Disease and pest control

Scientific farming pieces together the conditions and factors that make up production and either conducts specialized research in each area or arrives at generalizations, on the basis of which it attempts to increase yields.

The notion of raising productivity by making partial improvements in a number of these factors of production most likely originated with Liebig's thinking, which has played a key role in the development of modern agriculture in the West.

According to Liebig's law of minimum, the yield of a crop is determined by that nutrient present in shortest supply. Implicit in this rule is the notion that yield can be increased by improving the factors of production. And going one step further, this can also be understood to imply that, because the worst factor represents the largest barrier to increased yields, significant improvement can be made in the yield by training research efforts on this factor and improving it.

Using the analogy of a barrel (Fig. 2.5), Liebig's law states that, just as the level of the water in a barrel cannot rise above the height of the lowest barrel stave, so yields are determined by the factor of production present in shortest supply. In reality, however, this is not the case.

Granted, if we break down the crop nutrients and analyze them chemically, we find that these can be divided into any number of components: nitrogen, phosphorus, potassium, calcium, manganese, magnesium, and so on. But to claim that supplying all these factors in sufficient quantity raises yield is dubious reasoning at best. Rather than claiming that this increases yield, we should say only that it maintains yield. A nutrient in short supply decreases yield, but providing a sufficient amount of this nutrient does not increase yield, it merely prevents a loss in yield.

Liebig's barrel fails to apply to real-life situations on two counts. First, what holds up the barrel? The yield of a crop is not determined by just one factor; it is the general outcome of all the conditions and factors of cultivation. Thus, before becoming concerned with the effects that the surplus or shortage of a particular nutrient might have, it would make more sense to decide first the extent to which nutrients play a determining role on crop yields.

Unless one establishes the limits, coordinates, and domain represented by that factor known as nutrients, any results obtained from research on nutrients break apart in midair. Liebig's barrel is a concept floating on air. In the real world, yield is composed of innumerable interrelated factors and conditions, so the barrel should be shown on top of a column or pedestal representing these many conditions.

As Figure 2.7 shows, yield is determined by various factors and conditions, such as scale of operations, equipment, nutrient supply, and other considerations. Not only is the effect of a surplus or deficiency of any one factor on the yield very small, there is no real way of telling how great this effect is on a scale of one to ten.

Then also, the angle of the column or pedestal holding up the barrel affects the tilt of the barrel, changing the amount of water that it can hold. In fact, because the tilt of the barrel exerts a greater influence on the amount of water held by the barrel than the height of the staves, the level of individual nutrients is often of no real significance.

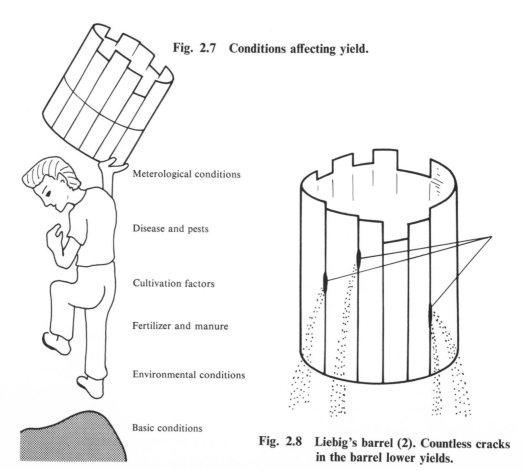

Fig. 2.7   Conditions affecting yield.

Meterological conditions

Disease and pests

Cultivation factors

Fertilizer and manure

Environmental conditions

Basic conditions

Fig. 2.8   Liebig's barrel (2). Countless cracks in the barrel lower yields.

The second reason Liebig's barrel analogy does not apply to the real world is that the barrel has no hoops. Before worrying about the height of the staves, we should look at how tightly they are fitted together. A barrel without hoops leaks horribly and cannot hold water. The leakage of water between the barrel staves due to the absence of tightly fastened hoops represents man's lack of a full understanding of the interrelatedness of different nutrients.

One could say that we know next to nothing about the true relationships between nitrogen, phosphorus, potassium, and the dozens of other crop nutrients; that no matter how much research is done on each of these, man will never fully understand the organic connections between all the nutrients making up a single crop.

Even were we to attempt to fully understand just one nutrient, this would be impossible because we would also have to determine how it is interrelated with all other factors, including soil and fertilizers, method of cultivation, pests, and the weather and environment. But this is impossible because time and space are in a constant state of flux. Not understanding the relationships between nutrients amounts to the lack of a hoop to hold the barrel staves together. This is the same as an agricultural test center with separate sections devoted to the study of cultivation techniques, fertilizers, and pest control; even the existence of a planning section and a farsighted director will be unable to pull these sections together into an integral whole with a common purpose.

The point of all this is simple: as long as Liebig's barrel is constructed of staves representing various nutrients, the barrel will not hold water. Such thinking cannot produce a true increase in yield. Examining and repairing the barrel will not raise the level of the water. Indeed, this can be done only by changing the very shape and form of the barrel.

Broad interpretation of Liebig's law of minimum leads to propositions such as "yield can be raised by improving each of the conditions of production," or "defective conditions being the controlling factors of yield, these should be the first to be improved." But these are equally untenable and false.

One often hears that yields cannot be increased in a certain locality because of poor weather conditions, or because soil conditions are poor and must first be improved. This sounds very much as if we were talking of a factory where production is the output of components such as raw materials, manufacturing equipment, labor, and capital. When a damaged gearwheel in a piece of machinery slows production in a factory, productivity can soon be restored by repairing the problem. But crop cultivation under natural conditions differs entirely from industrial fabrication in a plant. In farming, the organic whole cannot be enhanced by the mere replacement of parts.

Let us retrace the steps of agricultural research and examine the errors committed by the thinking underlying the law of minimum and analytical chemistry.

Research on crop cultivation began by examining actual production conditions. The goal being to increase production by improving each of these conditions, research efforts were divided initially into specialized disciplines such as tillage and seeding, soil and fertilizers, and pest control. As research progressed in each of these areas, the findings were collected together and applied by farmers to boost productivity. Factors identified as having a controlling influence on productivity were targeted as high-priority research topics.

Tillage and seeding specialists believe that improvements in these techniques are critical to increasing yields. They see such areas as when, where, and how to seed, and how to plow a field as the first topics research on crop cultivation should address.

A fertilizer specialist will tell you, "Keep fertilizing your plants and they'll just keep on growing. If it's high yields you're after, you've got to give your crops a lot of fertilizer. Increased fertilization is a positive way to raise yields." And the pest control specialist will say, "No matter how carefully you grow your crops and how high the yields you're after, if your fields are damaged by a crop disease or an insect pest, you're left with nothing. Effective disease and pest control is indispensable to high-yield production."

All such factors appear to help increase production, but the conventional view is that the tillage and seeding methods, breeding, and fertilizer application have a direct positive influence on yields, disease and pest damage reduce yields, and weather disasters destroy crops.

But are these actually important factors that work independently of each other under natural conditions to set or increase yields? And is there perhaps a range in the degree of importance of these factors? Let us consider natural disasters, which result in extensive crop damage.

Gales that occur when the rice is heading and floods shortly after transplantation can have a very decisive effect on yields regardless of the combination of production factors. However, the damage is not the same everywhere. The effects of a single gale can vary tremendously depending on the time and place. In a single stretch of fields, some of the rice plants will have lodged while others will remain standing; some heads of rice will be stripped clean, others will have less than a quarter of the grains remaining, and yet others more than three quarters remaining. Some rice plants submerged under flood waters will soon recover and continue growing, while others in the same waters will rot and die.

Damage may have been light because a host of interrelated factors—seed variety, method of cultivation, fertilizer application, disease and pest control—combined to give healthy plants that were able to recover as growth conditions and the environment returned to normal. Even inclement weather or a natural disaster is intimately and inseparately related with other production factors. So it is a mistake to think that any one factor can act independently to override all other factors and exert a decisive effect on yield.

This is true also for disease and pest damage. Twenty percent crop damage by rice borers does not necessarily mean a twenty percent decline in harvested grain.

Yields may actually rise in spite of pest damage. If a farmer expecting twenty percent crop damage by leafhoppers in his fields forgoes the use of pesticides, he may find the damage to be effectively contained by the appearance of vast numbers of spiders and frogs that prey on the leafhoppers.

Insect damage is the consequence of a number of causes. If we trace each of these back, we find that the damage attributable to any one cause is generally very insignificant. Natural farming takes a broad view of this tangle of causality and the interplay of different factors, and chooses to grow healthy crops rather than exercise pest control.

Breeding programs have sought to develop new high-yield strains that are easy to grow, resistant to insect pests and disease, and so on. But the creation and abandoning over the past several decades of tens of thousands of new varieties shows that the goals set for these change constantly, an indication that the question of seed variety cannot be resolved independently of other factors.

Although genetic enhancement techniques may be useful in achieving temporary gains in yield and quality, such gains are never permanent or universal. The same is true for methods of cultivation. Undeniable as it is that plowing a field, the time and period of seeding, and the raising of seedlings are basic to growing crops, we are wrong to think that the skill applied to these methods is decisive in setting yields.

Deep plowing was for a long time considered an important factor in determining crop yield, yet today a growing number of farmers no longer believe plowing to be necessary. Some even think that intertillage, weeding, and transplantation, all practices held to be of central importance by most farmers, are not needed at all. The use of such practices is dictated by the thinking of the times and other factors.

Another pitfall is the belief that fertilizers and methods of fertilizer application are directly linked to improved yields. Damage by heavy fertilization can just as easily lead to reduced yields. No single factor of production is powerful enough by itself to determine the yield or quality of a harvest. All are closely interrelated and share responsibility with many other factors for the harvest.

The moment that he applied discriminating knowledge to his study of nature, the scientist broke nature into a thousand pieces. Today, he picks apart the many factors that together contribute to the production of a crop, and studying each factor independently in specialized laboratories, writes reports on his research which he is confident, when studied, will help raise crop productivity. Such is the state of scientific agriculture today. While such research helps throw some light on current farming practices and may be effective in preventing a decline in productivity, it does not lead to discoveries of how to raise productivity and achieve spectacularly high yields.

Far from benefitting agricultural productivity, progressive specialization in research actually has the opposite effect. Methods intended to boost productivity lead instead to the devastation of nature, lowering overall productivity. Science labors under the delusion that the accumulated findings of an army of investigators pursuing research of limited scope in separate disciplines will provide a total and complete picture of nature.

Although parts may be broken off from the whole, "the whole is greater than the sum of the parts," as the saying goes. By implication, a collection of an infinite number of parts includes an infinite number of unknown parts. These may be represented as an infinite number of gaps, which prevent the whole from ever being completely reassembled.

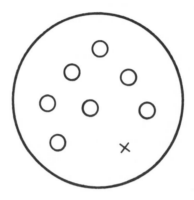

Fig. 2.9   The whole consists of known (○) and unknown (×) parts.

Scientific agriculture believes that by training specialized research on parts of the whole, partial improvements can be made which will translate into improvement of the whole. But nature should not forever be picked apart. Man has become so absorbed in his pursuit of the parts that he has abandoned his quest for the truth of the whole. Or perhaps, inevitably, his attempt to know the parts has made him lose sight of the whole.

Fragmented research only produces results of limited utility. All scientific farming can provide are partial improvements that may give high yields and increased production under certain conditions, but these "gains" soon fall victim to the violent recuperative backlash of nature and never ultimately result in higher yields.

Being limited and imperfect, human knowledge cannot hope to win out over the whole and ever-perfect wisdom of nature. Hence, all efforts to raise productivity founded on human knowledge can be successful only in a limited sense. While they may help deter a decline in yield by compensating for an irregular dip in productivity, such efforts will never be a means for significantly boosting productivity. Although man may interpret the result as an increase in yield, his efforts will never amount to anything more than a means for staving off reduced yields. All of which goes to show that, try as he may, man cannot equal the yields of nature.

Critique of the Inductive and Deductive Methods

Scientific thought is founded on inductive and deductive reasoning, so a critical review of these methods will allow us to examine the basic foundations of science. As my example, I will use the process of conducting research on rice cultivation.

One normally begins by drawing up a general proposition from a number of facts or observations. Thus, in the case of rice cultivation, a comprehensive study of rice is made. To determine the most suitable quantity of rice seed to be sown, the scientist experiments with a variety of seeding quantities. To establish the optimal spacing of plants, he runs tests in which he varies the number of days seedlings are grown in the nursery, and the number and spacing of transplanted seedlings. He compares several different varieties and selects those that give the highest yields. And to set guidelines for fertilizer application, he tries applying different amounts of nitrogen, phosphorus, and potassium. Inferences drawn from the results of these tests form the basis for selecting optimal techniques and quantities to be used in all methods of producing rice. The scientist or farmer, as the case may be, relies on these conclusions to make general decisions and erect standards that he believes help improve rice cultivation.

But do a number of disparate improvements add up to the best overall result? This problem lies behind the notable failure of most research to achieve higher yields in rice cultivation. Respective ten-percent improvements through new varieties of rice, tilling and seeding techniques, fertilization, and pest and disease control might be expected to add up to an overall increase of forty percent in yields, but actual improvements in the field amount to from 2 to 10 percent, at best.

Why do 1+1+1 not make 3, but 1? For the same reason that the pieces of a broken mirror can never be reassembled into a mirror more perfect than the original. The reason agricultural test stations were unable to produce more than 15–20 bushels per quarter-acre until around 1965 was that all they were doing, essentially, was analyzing and interpreting rice that already yielded 15–20 bushels per quarter-acre to begin with.

Although such research was launched to develop high-yielding techniques that are more productive than those used by the ordinary farmer, its only achievement has been the addition of scientific commentary on existing rice-growing methods. It has not improved on or exceeded farmer's yields. Such is the fate of inductive research.

Scientific agriculture first conducts research primarily by the inductive method, then does an about-face, applying deductive reasoning to draw specific propositions from general premises.

Natural farming arrives at its conclusions by applying deductive reasoning based on intuition. By this, I do not mean the imaginative formulation of wild hypotheses, but a mental process that attempts to reach a broad conclusion through intuitive understanding. During this process, it draws narrow conclusions adapted to the time and place, and searches out concrete ways in keeping with these conclusions.

Natural farming thus begins with conclusions and seeks concrete means of attaining these. This contrasts sharply with the inductive approach, which studies the current situation and from this derives a theory that it then uses to search for a conclusion while making gradual improvements along the way. In the first case, we have a conclusion, but no means of achieving it, and in the second, we have means at our disposal, but no conclusion.

Returning again to our example of rice cultivation, natural farming uses intuitive reasoning to draw up an ideal vision of rice cultivation, infers the environmental conditions under which a situation approximating the ideal can arise, and works out a means of achieving this ideal form of rice cultivation. On the other hand, scientific farming studies all aspects of rice production and conducts many different tests in an attempt to develop increasingly economical and high-yielding methods of rice cultivation.

Such inductive experimentation is done without a clear goal. Scientists run experiments oblivious to the direction in which their research takes them. They may be pleased with the results and confident that the amassing of such results leads to clear and steady progress and scientific achievement. But in the absence of a clear goal by which to set their course, their activity is just aimless wandering. It is not progress.

The scientist is well aware of the restrictive and circumstantial nature of inductive research, and does give some thought to deductive reasoning, but he ends up relying on the inductive approach because this leads more directly to practical and certain success and achievement.

Deductive experimentation has never had much appeal to scientists because they are unable to get a good handle on what appears to many a whimsical process. In addition, as this requires a great deal of time and space, it runs counter to the natural inclinations of scientists, who like to hole up in their laboratories. The reality is that both the inductive and the deductive method thread their way through the entire history of agricultural development. Of the two, deductive reasoning has always been the driving force behind rapid leaps in development, which are invariably triggered by some oddball idea thought up by an eccentric or a zealous farmer bit by curiosity.

Generally lacking scope and universality, this idea tends to slide back into oblivion unless the scientist recognizes it as a clue. After taking it apart and analyzing, studying, reconstructing, and verifying it through inductive experimentation, the scientist raises the idea to the level of a universally applicable technique. It is only at this point that the original idea is ready to be put to practical use and may, as often is the case, eventually become widely adopted by farmers.

Thus, although the guiding force of agriculture development is inductive reasoning by the scientist, the initial inspiration that lays the rails for progress is often the deductive notion of a progressive farmer or a hint left by someone who has nothing to do with farming.

Clearly then, the inductive method is useful only in a negative sense, as a means for preventing a decline in crop yields. Although throwing light on existing methods, it cannot break new ground in agriculture. Only deductive reasoning can bring forth fresh ideas having the potential of leading to positive gains in yields. Yet, because deductive reasoning generally remains poorly understood and is defined primarily in relation to induction, it is not likely to lead to any dramatic increases in yield.

True deduction originates at a point beyond the world of phenomena. It arises when one has acquired a philosophical understanding of the true essence of the natural world and grasped the ultimate goal. All that man sees is a projected image of nature. Unable to grasp the ultimate goal, he assumes deduction to be merely the inverse of induction and can go no further than deductive reasoning, which is but a dim shadow of true deduction. Experiments in which deduction is treated as merely the counterpart of induction have brought us the confusion of modern science. Even in agriculture, farmers and scientists are confounding measures for preventing crop losses with means for raising yields, and by discussing both on the same plane, are only prolonging the current stagnation of agriculture.

Induction and deduction can be likened to two climbers ascending a rock face. The lower of the two, who checks his footing before giving the climber in the lead a boost, plays an inductive role, while the lead climber, who lets down a rope and pulls the lower climber up, plays a deductive role.

Induction and deduction are complementary and together form a whole. Surprising as it may seem, although scientific agriculture has relied primarily on inductive experimentation, progress has been made as well in deductive reasoning. This is why measures to prevent crop losses and measures to boost yields have been confused.

Deduction here being merely a concept defined in relation to induction, we may see a gradual increase in yields, but are unlikely to see a dramatic increase. Our two climbers make only slow progress, and will never go beyond the peak they have already sighted.

To attain dramatically improved yields of a type possible only by a fundamental revolution in farming practices, one would have to rely not on this restricted notion of deduction, but on a broader deductive method that I will refer to here as "intuitive reasoning." In addition to our two climbers with a rope, other radically different methods of reaching the top of the mountain are possible, such as descending onto the peak by rope from a helicopter. It is from just such intuitive reasoning, which goes beyond induction and deduction, that the thinking underlying natural farming arises.

The creative roots of natural farming must be true intuitive understanding. The point of departure here must be a true understanding of nature gained by fixing one's gaze on the natural world that extends beyond actions and events in one's immediate surroundings. An infinitude of yield-improving possibilities lie hidden here. One must look intently at what lies beyond the immediate.

High-Yield Theory Is Full of Holes

It is all too easy for most people to think that scientific farming, which harnesses the forces of nature and adds to this human knowledge, is superior to natural farming both from the standpoint of economics and crop yields. This, of course, is not the case, for a number of reasons.

1. Scientific farming has isolated the factors responsible for yield and found ways to improve each of these. But although science can break nature down and analyze it, it cannot reassemble the parts into the same whole. What may appear to be nature reconstructed is just an imperfect imitation that can never produce higher yields than natural farming.

2. What is trumpeted as high-yield theory and technology amounts to nothing more than an attempt to approach natural harvests. Rather than aiming at large

jumps in yields, as is claimed, these are really just measures to stave off crop losses.

3. Not only does the endeavor to artificially achieve high yields that surpass natural output only increase the level of imperfection, it invites a breakdown in agriculture, and viewed in a larger sense, is just so much wasted effort. Yields that outstrip nature can never be achieved.

Fig. 2.10 Harvests compared.

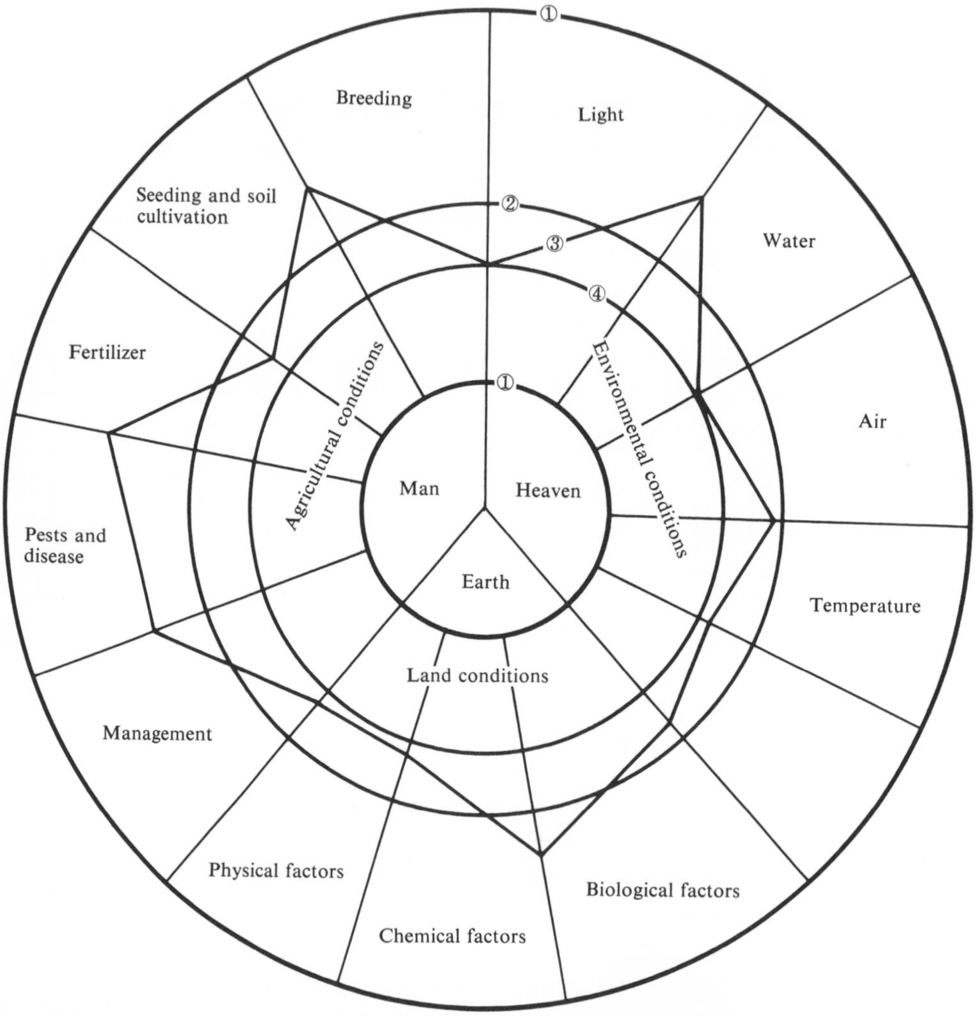

Circle ① represents the yields of Mahayana natural farming, circle ② the yields of Hinayana natural farming, circle ③ those of scientific farming, and circle ④ those based on Liebig's law.

The diagram in Fig. 2.10 compares the yields of natural farming and scientific farming. Outermost circle ① represents the yields of pure Mahayana natural farming. Actually, this cannot be properly depicted as either large or small, but lies in the world of Mu, shown as the innermost circle ① at the center of the diagram. Circle ② represents the yields of narrower, relativistic Hinayana natural farming.

Growth in these yields always parallels growth in the yields of scientific farming ③. Circle ④ stands for the yields likely to result from the application of Liebig's law of minimum.

A Model of Harvest Yields: A good way to understand how crop yields are determined by different factors or elements is to use the analogy of a building like that shown in Fig. 2.11. The hotel (or warehouse) is built on a rock foundation that symbolizes nature, and the floors and rooms of the building represent cultivation conditions and factors which play a role in the final yield. Each of the floors and rooms are integrally and inseparably related. This building shows us a number of things:

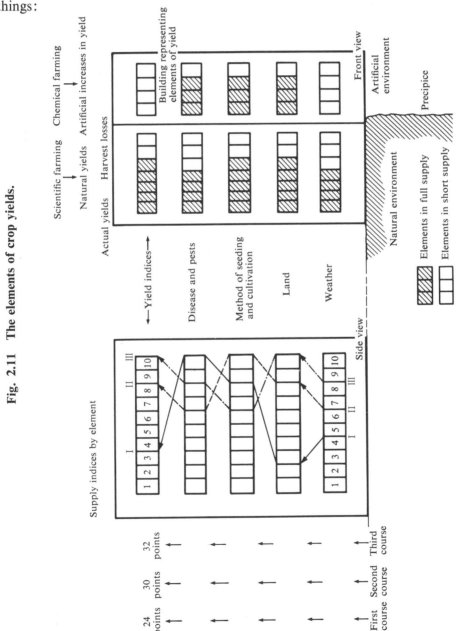

Fig. 2.11 The elements of crop yields.

1.   Yield is determined by the size of the building and the degree to which each room is full.

2.   The upper limit of yield is set by the natural environment, represented here by the strength of the rock foundation and the size of the building site. One could have gotten a reasonably close idea of the potential yield from the blue-prints of the building. The limit became fixed when the frame for the building was put into place. This maximum yield may be called the natural yield and is, for man, the best and highest yield.

3.   The actual harvest is much lower than this maximum yield, for the harvest does not completely fill each and every room. If the building were a hotel, this would be equivalent to saying that some guest rooms are vacant. In other words, there are invariably flaws or weaknesses in some of the elements of cultivation; these hold down yields. The actual harvest is what we are left with after sub-tracting the vacant rooms from the total number of guest rooms.

4.   The approach usually taken by scientific farming to boost yields is to fill as many of the rooms as possible. But in a larger sense, this is merely a way of minimizing losses in yield. The only true way to raise yields is to enlarge the building itself.

5.   Any attempt to outdo nature, to increase production by purely industrial methods that brazenly disregard the natural order, is analogous to adding an annex onto the building representing nature. If we imagine this annex to be built on sand, then we can begin to understand the precariousness of artificial en-deavors to raise yields. Inherently unstable, these do not represent true produc-tion and do not really benefit man.

6.   Although one would assume that filling each of the rooms would reduce losses and produce a net increase in yields, this is not necessarily so because all the rooms are closely interconnected. One cannot make selective improvements here and there in specific factors of production.

Knowing all this, we can better comprehend what the building signifies. To accept the thinking of Liebig is to say that yield is dominated by that element present in shortest supply. Such reasoning implies that, if one is not applying enough fertilizer or is using the wrong method of pest control, then correcting this will raise yields. Yet half-baked improvements of this sort are no more effec-tive than renovating just the fourth floor, or just one room on the first floor. The reason is that there are no absolute criteria with which to judge whether one ele-ment or condition is good or bad, excessive or insufficient. Such qualitative and quantitative characteristics of an element vary in a continuously fluid relationship with those of other elements; at times these work together, at other times they cancel each other out.

Being nearsighted, what man takes to be improvements in various elements are just localized improvements—like remodeling one room of the hotel. There is no way of knowing what effect this will have on the entire building.

One cannot know how business is faring at a hotel just by looking at the number of guest rooms or the number of vacancies. True, there may be many empty rooms, but other rooms may be packed full; in some cases, one good patron may be better for business than a large number of other guests. Good conditions in one room do not necessarily have a positive effect on overall business, and bad conditions on the first floor do not always exert a negative influence on the second and third floors. All the rooms and floors of the building are separate and distinct, yet all are intimately linked together into one organic whole. Although the claim can be made that the final yield is determined by the combination of an infinite array of factors and conditions, just as a new company president can dramatically change company morale, so the entire yield of a crop may turn on a change in a single factor.

In the final analysis, one cannot predict which element or factor will help or hurt the yield. This can only be determined by hindsight—after the harvest is in.

A farmer might say that this year's good harvest was due to the early-maturing variety he used, but he cannot be certain about this because of the unlimited number of factors involved. He has no way of knowing whether using the same variety the following year will again give good results.

One could even go to the extreme of saying that the effects of all the factors on the final yield can hinge, for example, on how a typhoon blows. This could turn bad conditions into good conditions. Last year's crop failure might have been the result of spreading too much fertilizer, which led to excessive plant growth and pest damage, but this year is windier so the fertilizer may succeed if the wind helps keep the bugs off the plants. We cannot predict what will work and what will not, so there is no reason for us to fret over minor improvements.

Just as the manager of our hotel will never succeed if all he pays attention to is whether the lights in the guest rooms are on or off, careful attention to tiny, insignificant details will never get the farmer off to a good start. Clearly, the only positive way to increase yields is to increase the capacity of the hotel. What we need to know is whether the hotel can be renovated, and if so, how.

We must not forget that as the scientist makes additions and repairs and the building gets higher and higher, it gets increasingly unstable and imperfect.

His observations, experiences, and ideas being entirely derived from nature, man can never build a house that extends beyond the bounds of nature, but heedless of this and not content with crops in their natural state, he has broken away from the natural arrangement of environmental factors and begun building an addition to the house of nature—artificially cultivated crops.

This artificial, chemically produced food unquestionably presents a dreadful danger to man. More than just a question of wasted effort and meaningless toil, it is the root of a calamity that threatens the very foundations of human existence. Yet agriculture continues to move rapidly toward the purely chemical and industrial production of agricultural crops, an addition—to return to my original analogy—built by man which projects out from the cliff serving as nature's foundation.

The side view of the building shows which path to follow in climbing from floor to floor while meeting the requirements for each of the factors of production.

For example, since Course I begins under poor weather and land conditions, the yield is poor regardless of special efforts invested in cultivation and pest control. Weather and land conditions in Course II are good, so the yield is high even though the method of cultivation and overall management leave something to be desired.

One cannot predict, however, which course will give the highest yield as there are an infinite number of courses, and infinite variations in each of the factors and conditions on these courses. While no doubt of use to the theorist for expounding the principles of crop cultivation, this diagram has no practical value.

A Look at Photosynthesis: Research aimed at high rice yields likewise begins by analyzing the factors underlying production. This commences with morphological observation, proceeds next to dissection and analysis, then moves on to plant ecology. By conducting laboratory experiments, pot tests, and small-scale field experiments under highly selective conditions, scientists have been able to pinpoint some of the factors that limit yields and some of the elements that increase harvests.

Yet clearly, any results obtained under such special conditions can say little about the incredibly complex set of natural conditions at work in an actual field. It comes as no surprise then that research is turning from the narrow, highly focused study of individual organisms to a broader examination of groups of organisms and investigations into the ecology of rice. One line of investigation being taken to find a theoretical basis for high yields is the ecological study of photosynthetic crops that increase starch production.

Many scientists continue to feel, however, that ecological research aimed at increasing the number of heads or grains of rice on a plant, or raising the size of the grains are crude and elementary. These people believe that physiological research which lays bare the mechanism of starch production is higher science; they subscribe to the illusion that such revelations will provide a basic clue to high yields.

To the casual observer, the study of photosynthesis within the leaves of the rice plant appears to be a topic of utmost importance, the findings of which could lead to a theory of high yields. Let us take a look at this research process. If one accepts that increasing the production of starch is connected to high yields, then research on photosynthesis does take on a great importance. And, as efforts are made to increase the amount of sunlight received by the plant and research is carried out on ways of improving the plant's capacity for starch synthesis from sunlight, people begin thinking that high yields are possible.

Current high-yield theory, as seen from the perspective of plant physiology, says essentially that yields may be regarded as the amount of starch produced by photosynthesis in the leaves of the plant, minus the starch consumed by respiration. Proponents of this view claim that yields can be increased by maximizing the photosynthetic ability of the plant while maintaining a balance between starch production and starch consumption.

But is all this theorizing and effort useful in achieving dramatic increases in rice yields? The fact of the matter is that today, as in the past, a yield of about 22 bushels per quarter-acre is still quite good, and the goal agronomists have set for themselves is raising the national average above this level. The possibility of reaping 26 to 28 bushels has recently been reported by some agricultural test centers, but this is only on a very limited scale and does not make use of techniques likely to gain wide acceptance and use. Why is it that such massive and persistent research efforts have failed to bear fruit? Perhaps the answer lies in the physiological processes of starch production by the rice plant and in the scientific means for enhancing the starch productivity of the plant.

Fig. 2.12 Starch production and consumption in a rice plant.

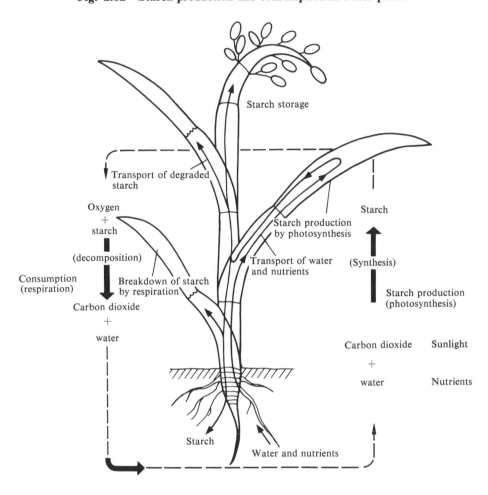

The diagram in Fig. 2.12 depicts a number of processes at work in the rice plant:
1) The leaves of the plant use photosynthesis to synthesize starch, which the leaves, stem, and roots consume in the process of respiration.
2) The plant produces starch by taking up water through the roots and sending it to the leaves, where photosynthesis is carried out using carbon dioxide absorbed through the leaf stomata and sunlight.
3) The starch produced in the leaves is broken down to sugar, which is sent to all parts of the plant and further decomposed by oxidation. This degradative process of respiration releases energy that feeds the rice plant.

4) A large share of the starch produced in this way is metabolized by the plant and the remainder stored in the grains of rice.

Armed with a basic understanding of how photosynthesis works, the next thing science does is study ways in which to raise starch productivity and increase the amount of stored starch. Innumerable factors affect the relative activities of photosynthesis and respiration. Here are some of the most important:

Factors affecting photosynthesis: carbon dioxide, stomata closure, water uptake, water temperature, sunlight.

Factors affecting respiration: sugar, oxygen, strength of wind, nutrients, humidity.

One way of raising rice production that immediately comes to mind here is to maximize starch production by increasing photosynthesis while at the same time holding starch consumption down to a minimum in order to leave as much unconsumed starch as possible in the heads of rice.

Conditions favorable for high photosynthetic activity are lots of sunlight, high temperatures, and good water and nutrient uptake by the roots. Under such conditions, the leaf stomata remain open and much carbon dioxide is absorbed, resulting in active photosynthesis and maximum starch synthesis.

There is a catch to this, unfortunately. The same conditions that are favorable to photosynthesis also promote respiration. Starch production may be high, but so is starch consumption, and so these conditions do not result in maximum starch storage. On the other hand, a low starch production does not necessarily mean that yields will be low. In fact, if starch consumption is low enough, the amount of stored starch may even be higher—meaning higher yields—than under more vigorous photosynthetic activity.

How often have farmers and scientists tried techniques that maximize starch production only to find the result to be large rice plants that lodge under the slightest breeze? A much easier and more certain path to high yields would be to hold down respiration and grow smaller plants that consume less starch. The combinations of production factors and elements that can occur in nature are limitless and may lead to any number of different yields.

Various courses are possible in Fig. 2.13. For example, when there is abundant sunlight and temperatures are high—around 40°C (104°F), as in Course 1, root rot tends to occur, reducing root vitality. This weakens water uptake, causing the plant to close its stomata to prevent excessive loss of water. As a result, less carbon dioxide is absorbed and photosynthesis slows down, but because respiration continues unabated, starch consumption remains high, resulting in a low yield.

In Course 2, temperatures are lower—perhaps 30°C (86°F), and better suited to the variety of rice. Nutrient and water absorption are good, so photosynthetic activity is high and remains in balance with respiration. This combination of factors gives the highest yield.

In Course 3, low temperatures prevail and the other conditions are fair but hardly ideal. Yet, because good root activity supplies the plant with ample nutrients, a normal yield is maintained.

Fig. 2.13   There are many paths to a harvest.

Yields

| low | moderate | high | Finish line |

Respiration | weak | average | strong | Photosynthesis

Sugar | low | average | high | Carbon dioxide

Oxygen | low | average | high | Stomata closure

Wind strength | weak | moderate | strong | Water uptake

Nutrients | low | moderate | high | Temperature

Humidity | low | average | high | Sunlight

Starting line

Course

Small rice plants   Tug of war   Large rice plants

This is just a tiny sampling of the possibilities, and I have made only crude guesses at the effects several factors on each course might have on the final yield.

But in the real world yields are not determined as simply as this. An infinite number of courses exist, and each of the many elements and conditions during cultivation change, often on a daily basis, over the entire growing season. This is not like a footrace along a clearly marked track that begins at the starting line and ends at the finish line.

Even were it possible to know what conditions maximize photosynthetic activity, one would be unable to design a course that assembles a combination of the very best conditions. The best conditions cannot be combined under natural circumstances. And to make matters even worse, maximizing photosynthesis does not guarantee maximum yields; nor do yields necessarily increase when respiration is minimized. To begin with, there is no standard by which to judge what "maximum" and "minimum" are.

One cannot flatly assert, for example, that 40°C is the maximum temperature, and 30°C optimal. This varies with time and place, the variety of rice, and the method of cultivation. We cannot even know for certain whether a higher temperature is better or worse.

Another reason why we cannot know is that the notion of what is appropriate differs for each condition and factor. People are usually satisfied with an optimal temperature that is workable under the greatest range of conditions. Although this answers the most common needs and will help raise normal yields, it is not the temperature required for high yields. Our inquiry into what temperatures are needed for high yields thus proves fruitless and we settle in the end for normal temperatures.

What about sunlight? Sunlight increases photosynthesis, but an increase in sunlight is not necessarily paralleled by a rise in yields. In Japan, yields are higher in the northern part of Honshu than in sunny Kyushu to the south, and Japan boasts better yields than countries in the southern tropics. Everyone is off in search of the optimal amount of sunlight, but this varies in relation with many other factors.

Good water uptake envigorates photosynthesis, but flooding the field can hasten root decay and slow photosynthesis. A deficiency in soil moisture and nutrients may at times help to maintain root vigor, and at others may inhibit growth and bring about a decline in starch production. It all depends on the other conditions.

An understanding of rice plant physiology can be applied to a scientific inquiry into how to maximize starch production, but this will not be directly applicable in practice. Scientific visions of high yields based on the physiology of the rice plant amount to just a lot of empty theorizing. Maybe the numbers add up on paper, but no one can build a theory like this and get it to work in practice. The rice scientist well-versed in his particular specialty is not unlike the sports commentator who can give a good rundown of a tennis match and may even make a good coach, but is not a top-notch athlete.

This inability of high-yield theory to translate into practical techniques is a basic inconsistency that applies to all scientific theory and technology. The scientist is a scientist and the farmer a farmer and "never the twain shall meet." The scientist may study farming, but the farmer can grow crops without knowing anything about science. This is borne out nowhere better than in the history of rice cultivation.

Look Beyond the Immediate Reality: Obviously, productivity and yields are measured in relative terms. A yield is high or low with respect to some standard. In seeking to boost productivity, we first have to define a starting point relative to which an increase is to be made. But do not we in fact always aim to produce more, to obtain higher yields, while believing all the while that no harm can come of simply moving ahead one step at a time.

When people discuss rice harvests, for some reason they are usually most concerned with attempts to increase yields. By "high-yielding" all we really mean is higher than current rice yields. This might be 20 bushels per quarter-acre in some cases, and over 25 bushels in others. There is no set target for "high-yielding" cultivation.

The point of departure defines the destination, and starting blocks make sense only when there is a finish line. Without starting blocks we cannot take off. So it is meaningless to talk of great or small, gain or loss, good or bad.

Because we take the present for granted as certain and unquestionable reality, we normally make this our point of departure and view as desirable any conditions or factors of production that improve on it. Yet the present is actually a very shaky and unreliable starting point because a good hard look at this so-called reality shows the greater part of it to be man-made, to be erected on common-sensical notions, with the stability of a building erected on a ship.

Taking any one of the traditional notions of rice cultivation—plowing, starter beds, transplantation, flooded paddies—as our basic point of departure would be a grievous error. Indeed, true progress can be had only by starting out from a totally new point.

But where is one to search for this starting point? I believe that it must be found in nature itself. Yet philosophically speaking, man is the only being that does not understand the true state of nature. He discriminates and grasps things in relative terms, mistaking his phenomenological world for the true natural world. He sees the morning as the beginning of a new day; he takes germination as the start in the life of a plant, and withering as its end. But this is nothing more than biased judgment on the part of man.

Nature is one. There is no starting point or destination, only an endless flux, a continuous metamorphosis of all things. Even this may be said not to exist. The true essence of nature then is "nothingness." It is here that the real starting point and destination are located. To make nature our foundation is to begin at "nothing" and make this point of departure our destination as well, to start off from "nothing" and return to "nothing," to make nature both the starting point and terminus of our journey. We should not make conditions directly before us a platform from which to launch new improvements. Instead, we should distance ourselves from the immediate situation, and observing it at a remove, at a point of Mu, or "nothingness," should aim to return to the Mu of nature.

This may seem to be very difficult, but may also appear very easy because the world beyond immediate reality is actually nothing more than the world as it was prior to human awareness of reality. A look from afar at the total picture is no better than a look up close at a small part because both are one inseparable whole. This undivided and inseparable unity is the "nothingness" that must be understood as it is. To start from nothing and return to nothing, that is natural farming.

If we strip away the layers of human knowledge and action from nature one by one, true nature will emerge of itself. A good look at the natural order thus revealed will show us just how great have been the errors committed by science. A science that rejects the science of today will surely ensue. Crops need only be entrusted to the hand of nature. The starting point of natural farming is also its destination, and the journey in-between.

One may believe the productivity of nature farming—which has no notion of time and space—to be quantifiable or unquantifiable; it makes no difference. Natural farming merely provides harvests that follow a fixed, unchanging orbit with the cycles of nature. Yet, let there be no mistake about it, natural harvests always give the best possible yields and are never inferior to the harvests of scientific farming.

The scientific world of "somethingness" is smaller than the natural world of "nothingness." No degree of expansion can enable the world of science to arrive at the vast, limitless world of nature.

Original Factors Are Most Important: We have seen that resolving production into elements or constituent factors and studying ways of improving these individually is basically an invalid approach. Now I would like to examine the propriety of scientists ignoring correlations between different factors, of their adherence to a sliding scale of importance in factors, and of their selective study of those elements that offer the greatest chances for rapid and visible improvement in yields.

The factors involved in production are infinite in number, and all are organically interrelated. None exerts a controlling influence on production. Moreover, these cannot and should not be ranked by importance.

Each factor is meaningful in the tangled web of interrelationships, but ceases to have any meaning when isolated from the whole. In spite of this, individual factors are extracted and studied in isolation all the time. Which is to say that research attempts to find meaning in something from which it has wrested all meaning.

There are commonly thought to exist a number of important topics that should be addressed, factors that should be studied, in order to be able to raise crop production. Since people feel that the quickest way to raise production is to make improvements in those factors thought to be deficient in some way (Liebig's law of minimum), they sow seed, apply fertilizer, and control disease and insect damage. So it comes as no surprise when research follows suit by focusing on methods of cultivation, soil and fertilizers, disease and insect pests. Environmental factors such as climate that are far more difficult for man to alter are given a wide berth.

But judging from the results, the factors most critical to yields are not those which man believes he can easily improve, but rather the environmental factors abandoned by man as intractable. Moreover, it is precisely those factors that we break down, meticulously categorize, and view as vital and important that are the most trivial and insignificant. Those primitive, unresolved factors not yet subjected to the full scrutiny of scientific analysis are the ones of greatest importance.

The fact that agricultural test centers are divided into different sections such as breeding, cultivation, soil and fertilizers, and plant diseases and pests is proof that agricultural research does not take a comprehensive approach to the study of nature. Instead, it starts from simple economic concerns and proceeds wherever man's desires take him, with the result that fragmented research is conducted in response to the concerns of the moment, almost as if by impulse.

Whichever field of inquiry we look at—plant breeders who chase after rare and unusual strains; agronomics and its preoccupation with high yields; soil science based on the premise of fertilizer application; entomologists and plant pathologists who devote themselves entirely to the study of pesticides for controlling diseases and pests without ever giving a thought to the role played by poor plant health; and meteorologists who perform token research in agricultural meteorology, a marginal and very narrowly defined discipline that only gets any attention when there is no other alternative—one thing is clear: modern agricultural research is not an attempt to gain a better understanding of the relationship between agricultural crops and man. From beginning to end, this has consisted exclusively of limited, inconsequential analytic research on single crops that does not set as its goal an understanding of the interrelationships between man and crops in nature.

As research grows increasingly specialized, it advances into ever more narrowly defined disciplines and penetrates into ever smaller worlds. The scientist believes that his studies reach down to the deepest stratum of nature, and his efforts bring man that much closer to a fundamental understanding of the natural world, but these endeavors are just peripheral research that moves further and further away from the fountainhead of nature.

Early man rose with the sun and slept on the ground. In ancient times, the rays of the sun, the soil, and the rains raised the crops; people learned to live by this and were grateful to the heavens and earth.

The man of science is well versed in small details and confident that he knows more about growing crops than the farmer of old. But does the scientist—who is aware that starch is produced from carbon dioxide and water by photosynthesis within the leaf with the aid of chlorophyll, and that the plant grows with the energy released by the oxidation of this starch—know more about light and air than the farmer who thinks the rice has ripened by the blessings of the sun? Certainly not! The scientist knows only one aspect, only one function of light and air—that seen from the perspective of science. Unable to perceive light and air as broadly changing phenomena of the universe, man isolates these from nature and examines them in cross-section like dead tissue under a microscope. In fact, the scientist, unable to see light as anything other than a purely physical phenomenon, is blind to light.

The soil scientist explains that crops are not raised by the earth, but grow under the effects of water and nutrients, and that high yields can be obtained when these are applied at the right time in the proper quantity. But he should also know that the soil he has in his laboratory is dead, mineral soil, not the living soil of nature. He should know that the water that flows down from the mountains and into the earth differs from the water that runs over the plains as a river; that the fluvial waters that give birth to all forms of life, from microorganisms and algae to fish and shellfish, are more than just a compound of oxygen and hydrogen.

Farmers build greenhouses and hot beds and grow vegetables and flowers there without knowing what sunlight really is or bothering to take a close look at how light changes when it passes through glass or vinyl sheeting. No matter how high a market price they fetch, the vegetables and flowers grown in such enclosures cannot be truly alive or of any great value.

No Understanding of Causal Relationships: The farmer might talk about how this year's poor harvest was due to the poor weather, while the specialist will go into more detail: "Tiller formation was good this year, producing a large number of heads. The grain count per head was also good, but insufficient sunlight after heading slowed maturation, resulting in a poor harvest."

The second explanation is far more descriptive and appears closer to the real truth. Surely one reason for poor maturation is insufficient sunshine, since the two clearly are causally related. Yet one cannot make the claim that poor sunlight during heading was the decisive factor behind the poor harvest that year, nor can one say that the poor harvest was due to a lack of sunlight.

This is because the causal relationship between these two factors—maturation and sunlight—is unclear. Poor maturation and insufficient sunlight mean that not enough sunlight was received by the leaves. The cause for this may have been drooping of the leaves due to excessive vegetative growth, and the drooping may have been caused by any number of factors. Perhaps this was a result of the overapplication and absorption of nitrogenous fertilizers, or a shortage of some other nutrient. Perhaps the cause was stem weakness due to a deficiency of silicate, or maybe the leaf droop was caused merely by an excess of leaf nitrogen on account of inhibition, for some reason, of the conversion of nitrogenous nutrients to protein. Behind each cause lies another cause.

When we talk of causes, we refer to a complex web of organically interrelated causes—basic causes, remote causes, contributing factors, predisposing factors. This is why one cannot give a brief, simple explanation of the true cause of poor maturation, and it is also why a more detailed explanation is no closer to grasping the real truth.

The poor harvest might be attributed to insufficient sunshine or to excess nitrogen during heading or merely to poor starch transport due to inadequate water. Or perhaps the basic cause is low temperatures. In any event, it is impossible to tell what the real cause is.

So what do we do? The conclusion we draw from all this is that the poor harvest resulted from a combination of factors, which is no more meaningful than the farmer saying it was written in the stars. The scientist may be pleased with himself for coming up with a detailed explanation, but it makes not the slightest bit of difference whether we carefully analyze the reasons for the poor harvest or throw all analysis to the winds; the result is the same.

Scientists think otherwise, however, believing that an analysis of one year's harvest will benefit rice growers the following year. However, the weather is never the same, so the rice growing environment next year will be entirely different from this year's. And because all factors of production are organically interrelated, when one factor changes, this effects changes in all other factors and conditions. This means that rice will be grown under entirely different conditions next year, rendering this year's experience and observations totally useless. Although useful for examining results in retrospect, the explanations of yesterday cannot be used to set tomorrow's strategy.

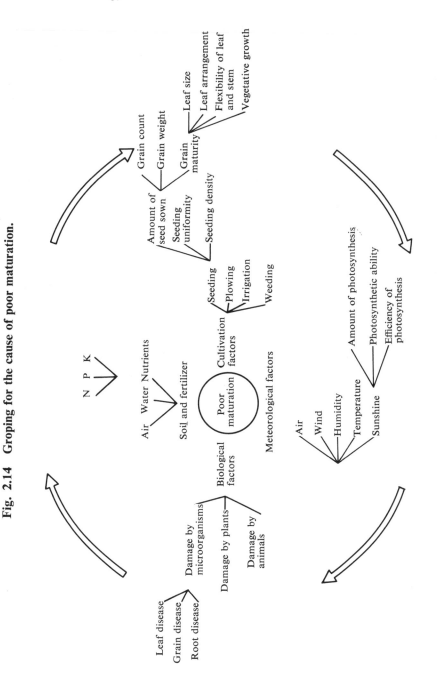

Fig. 2.14 Groping for the cause of poor maturation.

The causal relationships between factors in nature are just too entangled for man to unravel through research and analysis. Perhaps science succeeds in advancing one slow step at a time, but because it does so while groping in total darkness along a road that goes on forever, it is unable to know the real truth of things. This is why scientists are pleased with partial explications and see nothing wrong with pointing a finger and proclaiming this to be the cause and that the effect.

The more research progresses, the larger the body of scholarly data grows. The antecedent causes of causes increase in number and depth, becoming incredibly complex, such that, far from unraveling the tangled web of cause and effect, science succeeds only in explaining in ever greater detail each of the bends and kinks in the individual threads. There being infinite causes for an event or action, there are infinite solutions as well, and these together deepen and broaden to infinite complexity.

To resolve the single matter of poor maturation, one must be prepared to resolve at the same time all elements in every field that bear upon this—such as weather, the biological environment, the method of cultivation, soil, fertilizer, disease and pest control, and human factors. A look at the prospects of such a simultaneous solution should be enough to make man aware of just how difficult and fraught with contradiction this endeavor is. Yet, in a sense, this is already unavoidable.

Many people think that if you take a variety of rice that bears large heads of grain, grow it so that it receives lots of sunlight, apply plenty of fertilizer, and carry out thorough pest control measures, you will get good yields. However, varieties that bear large heads usually have fewer heads per plant. Thus it will do no good to plant densely if the intention is to allow better exposure to sunlight. Moreover, the heavy application of fertilizers will cause excessive vegetative growth, again defeating attempts to improve exposure to sunlight. Efforts to obtain large stems and heads only weaken the rice plant and increase disease and insect damage, while thorough pest control measures cause the rice plants to lodge.

The use of water-conserving rice cultivation to improve light exposure of the rice plants may actually cut down the available light due to the growth of weeds, and the lack of sufficient water may even interfere with the transport of nutrients. An attempt to raise the efficiency of photosynthesis may lower the photosynthetic ability of the plant. Apparently then, we conclude, irrigation is beneficial for the rice plants. So we try irrigating, and just when high temperatures would be expected to encourage vigorous growth, root rot sets in, resulting in poor maturation.

In other words, while a means of improving photosynthesis may prove effective at increasing the amount of starch, it does not necessarily exert a beneficial influence on those other elements that help set harvest yields and is in fact more likely to have countless negative effects.

In short, there is no way to combine all these into one overall method that works just right. The more improvement measures are combined, the more these measures cancel each other out to give an indefinite result, so that the only conclusion ends up being no clear conclusion at all.

If what people have in mind is that a plant variety that bears in abundance, is easy to raise, and has a good flavor would solve everything, they are in for a long wait. The day will never come when one variety satisfies all conditions.

The breeding specialist may believe that his endeavors will produce a variety that meets the needs of his age, but an improved variety with three good features will also have three bad features, and one with six strengths will have six weaknesses. All of which goes to show that any variety thought to be better will probably be worse, because in it will lie new contradictions that defy solution.

Although when examined individually, each of the improvements conceived by agricultural scientists may appear fine and proper, when seen as a whole they cancel each other out and are totally ineffective.

This property of mutual cancellation derives from the equilibrium of nature. Nature inherently abhors the unnatural and makes every effort to return to its true state by discarding human techniques for increasing harvests. For this reason, a natural control operates to hold down large harvests and raise low harvests, such as to approach the natural yield without disrupting the balance of nature.

In any case, since the basic causes of actions and effects that arise at any particular time and place cannot be known to man, and he can have no true understanding of the causal relationships involved, then there is no way for him to know the true effectiveness of any of his techniques. Although he knows that in the long run no conclusion is forthcoming, man persists nevertheless in the belief that his temporary conclusions and devices are effective in an overall sense. It is utterly impossible to predict what effects will arise from actions undertaken with the human intellect. Man only thinks the effects will be beneficial. He cannot know.

Although it would be desirable to erect overall measures and simultaneously apply methods complete on all counts, only God is capable of doing this. As the correlations and causal relationships between all the elements of nature remain unclear, man's understanding and interpretation can at best be only myopic and uncertain. His efforts thus cancel each other out and, after having succeeded only in causing meaningless confusion, are eventually buried in nature.

The Theory of Natural Farming

1. The Relative Merits of Natural Farming and Scientific Agriculture

Two Ways of Natural Farming

Although I have already shown in some detail the differences between natural farming and scientific farming, I would like to return here to compare the principles on which each is based. For the sake of convenience, I shall divide natural farming into two types and consider both.

Mahayana Natural Farming: When the human spirit and human life blend with the natural order and man's sole calling is to serve nature, he lives freely as an integral part of the natural world, subsisting on its bounty without having to resort to purposeful human effort. This type of farming, which I shall call Mahayana natural farming, is realized when man becomes one with nature, for it is a way of farming that transcends time and space and reaches the zenith of understanding and enlightenment.

This relationship between man and nature is like an ideal marriage in which the partners together realize a perfect life without asking for, giving, or receiving anything of each other. Mahayana farming is the very embodiment of life in accordance with nature. Those who live such a life are hermits and wise men.

Hinayana Natural Farming: This type of farming arises when man earnestly seeks entry to the realm of Mahayana farming. Desirous of the true blessings and bounty of nature, he prepares himself to receive it. This is the road leading directly to complete enlightenment, but is short of that perfect state. The relationship between man and nature here is like that of a lover who yearns after his loved one and asks for her hand, but has not realized full union.

Scientific Farming: Man exists in a state of contradiction in which he is basically estranged from nature, living in a totally artificial world, yet longs for a return to nature. A product of this condition, scientific farming forever wanders blindly back and forth, now calling upon the blessings of nature, now rejecting it in favor of human knowledge and action. Returning to the same metaphor, our lover here is unable to decide whose hand to ask in marriage, and while agonizing over his indecision, imprudently courts the maidens, heedless of social proprieties.

Absolute World	*Mahayana natural farming* (philosopher's way of farming)=pure natural farming
Relative World	*Hinayana natural farming* (idealistic farming)= natural farming, organic farming
	Scientific farming (dialectical materialism)= scientific agriculture

The Three Ways of Farming Compared: These may be arranged as before or depicted in the manner shown in Fig. 3.1.

Fig. 3.1 Three ways of farming.

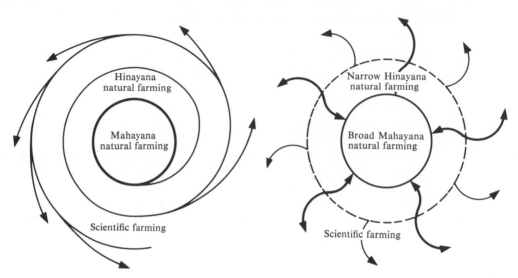

1. *Mahayana natural farming* and scientific farming are on entirely different planes. Although it is a bit strange to directly compare the two and discuss their relative merits, the only way we have of expressing their value in this world is by comparison and contrast. Scientific agriculture draws as much as it can from natural forces and attempts, by adding human knowledge, to produce results that eclipse nature. Naturally, proponents of this type of farming think it superior to natural farming, which relies on the forces and resources of nature.

Philosophically, however, scientific farming cannot be superior to Mahayana natural farming because, while scientific farming is the sum of knowledge and forces extracted from nature by human intellect, this still amounts to finite human knowledge. No matter how one totals it up, human knowledge is but a tiny, closely circumscribed fraction of the infinitude of the natural world. In contrast to the vast, boundless, and perfect knowledge and power of nature, the finite knowledge of man is always limited to small pockets of time and space. Inherently imperfect as it is, human knowledge can never be collected together to form perfect knowledge.

As imperfection can never be the equal of perfection, so scientific farming must always surrender a step to Mahayana natural farming. Nature has everything and, no matter how desperately he struggles, man will never be more than a small, imperfect part of its totality. Clearly then, scientific farming, by nature incomplete, can never hope to attain the immutable absoluteness of natural farming.

2. *Hinayana natural farming*, however, belongs in the same world of relativity as scientific farming, and so the two may be directly compared. Both are alike in that they are derived from nature and are verified with discriminating knowledge, but Hinayana farming attempts to cast off human knowledge and action and devote itself to making the greatest possible use of the pure forces of nature, whereas scientific farming uses the powers of nature and adds human knowledge and action in an effort to establish a superior way of farming.

The two differ fundamentally and are diametrically opposed in their perceptions, thinking, and research, but to explain the methods of Hinayana farming, we have no choice but to borrow the terms and methods of science. So for the sake of simplicity, we shall place it temporarily in the realm of science. In this respect, it resembles the position of the Eastern arts of healing vis-a-vis Western medicine. The direction in which Hinayana natural farming is headed takes it beyond the world of science and to a rejection of scientific thinking.

Borrowing a metaphor from the art of swordfighting, Hinayana natural farming may be likened to the one-sword school that is directed towards the center, and scientific farming to the two-sword school that is directed outwards. The two can be compared. But Mahayana natural farming is the unmoving no-sword school, comparison with which is impossible. Scientific farming uses all possible means at its disposable, increasing the number of swords, while natural farming tries to obtain the best possible results while rendering all means useless, in effect reducing the number of its swords (Hinayana) or doing entirely without (Mahayana).

This view is based on the philosophical conviction that if man makes a genuine effort to approach nature, then even should he abandon all deeds and actions, nature will take each of these over and perform them for him.

3. *Scientific farming:* Pure natural farming should therefore be judged on philosophical grounds, while scientific farming should be evaluated on a scientific basis. Because scientific farming is limited to direct circumstances in every respect, its achievements may excel in a restricted sense but are invariably inferior in all other ways. In contrast, natural farming is total and comprehensive, so its achievements must be judged from a broad, universal perspective.

When scientific methods are used to grow a fruit tree, for example, the goal may be to produce large fruit, in which case all efforts will be concentrated in this direction. Yet all that will be achieved is the production of what may, in a limited sense, be regarded as large fruit. The fruit produced by scientific farming is always large—even unnaturally so—in a relative sense, but invariably has grave flaws. Essentially, what is being grown is deformed fruit. To determine the true merit of scientific farming, one has to decide whether producing large fruit is truly good for man. The answer to this should be obvious.

Scientific farming constantly practices the unnatural without the slightest concern, but this is of very great significance and invites the gravest of consequences. The unnaturalness of scientific farming leads directly to imperfection, which is why its results are always distorted and at best of only local utility.

As the diagram in Fig. 3.2 shows, scientific farming and Hinayana natural farming both occupy the same dimension and may be described as the inner and outer edges of one circle, although one large difference is the very irregular contour of scientific farming.

The irregular shape of scientific farming represents the distortions and imperfections arising from the collection of narrow and limited research results of which it is made. This contrasts sharply with the perfect circle that signifies the perfection of nature towards which Hinayana natural farming aspires.

Because the nature seen by man is just a projected image of true nature, the circle representing Hinayana farming is drawn much smaller than that for Mahayana natural farming. Mahayana farming, which is nature itself, is better in every respect than the other ways of farming.

Fig. 3.2 Mahayana natural farming is absolute and beyond comparison.

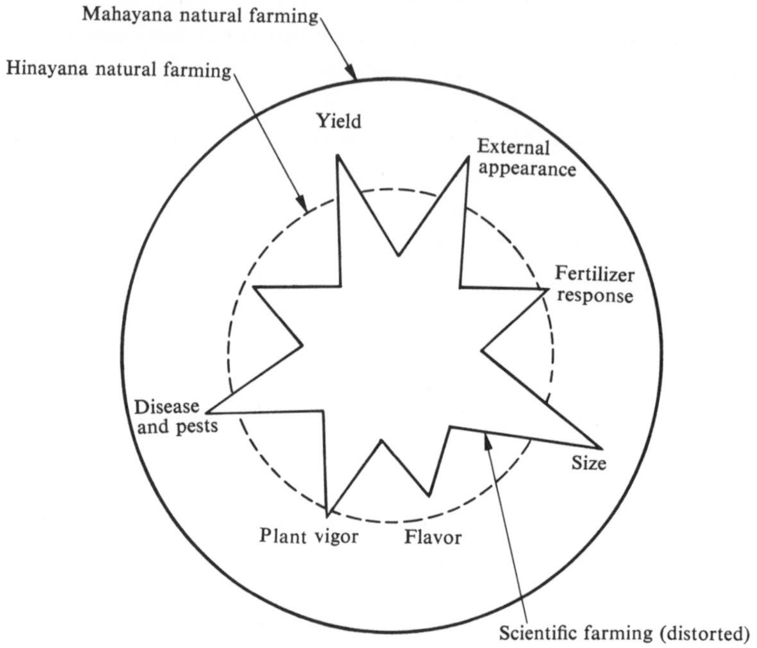

Scientific Agriculture: Farming Without Nature

Constant shifts and changes in crop growing practices and the history of sericulture and livestock farming show that while man may have approached natural farming in some ages, he leaned more towards scientific agriculture in others. Farming has repeatedly turned back to nature, then moved away again. Today, it is headed toward fully automated and systematized production.
The immediate reason for this trend toward mechanized agriculture is that artificial methods of raising livestock and scientific crop cultivation are believed to give higher yields and to be more economically advantageous, meaning higher productivity and profits.

Natural farming, on the other hand, seems but a passive and primitive way of farming, at best a laissez-faire form of extensive agriculture that gives meager harvests and paltry profits.

Here is how I compare the yields for these three types of farming:

1) Scientific farming excels under unnatural, man-made conditions. But this is only because natural farming cannot be practiced under such conditions.
2) Under conditions approaching those of nature, Hinayana natural farming will yield results at least as good as or better than scientific farming.
3) In wholistic terms, Mahayana natural farming, which is both pure and perfect, is always superior to scientific farming.

Let us take a look at situations in which each of these excel.

1. *Cases where scientific farming excels:* Scientific methods will always have the upper hand when growing produce in an unnatural environment and under unnatural conditions that deny nature its full powers, such as accelerated crop growth and cultivation in cramped plots, clay pots, hothouses, and hotbeds. And through adroit management, yields can be increased and fruit and vegetables grown out of season to satisfy consumer cravings by pumping in lots of high technology in the form of chemical fertilizers and powerful disease and pest control agents, bringing in unheard-of profits. Yet this is only because under such unnatural conditions natural farming does not stand a chance.

Instead of being satisfied with vegetables and fruit ripened on the land under the full rays of the sun, people vie with each other to buy pale, limpid, out-of-season vegetables and splendid-looking fruit packed with artificial coloring the minute these appear in the supermarkets and food stalls. Under the circumstances, it is no surprise that people are grateful for scientific farming and think of it as beneficial to man.

Yet even under such ideal conditions, scientific farming does not produce more at lower cost or generate higher profits per unit area of land or per fruit tree than natural farming. It is not economically advantageous because it produces more and better product with less work and at lower cost. No, it is suited rather to the skillful use of time and space to create profit.

People construct buildings on high-priced land and raise silkworms, chickens, or hogs. In the winter they grow tomatoes and watermelons hydroponically in large hothouses. Mandarin oranges, which normally ripen in late autumn, are shipped from refrigerated warehouses in the summer and sold at a high profit. Here scientific agriculture has the entire field to itself. The only response possible to a consumer public that desires what nature cannot give it is to produce crops in an environment divorced from nature and to allow science and technology that relies on human knowledge and action to flex its muscle.

But I repeat, viewed in a larger sense that transcends space and time, scientific farming is *not* more economical or productive than natural farming. This superiority of scientific farming is a fragile, short-lived thing, and soon collapses with changing times and circumstances.

2. *Cases where both ways of farming are equally effective:* Which of the two approaches is more productive under nearly natural conditions such as field cropping or the summer grazing of livestock? Under these circumstances, natural farming will never produce results inferior to scientific agriculture because it is able to take full advantage of nature's forces.

The reason is simple: man imitates nature. No matter how well he thinks he knows rice, he cannot produce it from scratch. All he does is take the rice plant that he finds in nature and try growing it by imitating the natural processes of rice seeding and germination. Man is no more than a disciple of nature. It is a foregone conclusion that, were nature—the teacher—to use its full powers, man—the disciple—would lose out in the confrontation.

A typical response might go as follows: "But a student sometimes catches up with and overtakes his teacher. Isn't it possible that man may one day succeed in fabricating an entire fruit. Even if this is not identical to a natural fruit, but a mere copy, might not it possibly be better than the real thing?"

But has anyone actually given any thought as to how much scientific knowledge, to the materials and effort, it would take to reproduce something of nature? The level of technology that would be needed to create a single persimmon seed or leaf is incomparably greater than that used to launch a rocket into outer space. Even were man to undertake a solution to the myriad mysteries in a persimmon seed and attempt to fabricate a single seed artificially, the world's scientists pooling all their knowledge and resources would not be up to the task.

And even were this possible, if man then set his mind on replacing current world fruit production with fruit manufactured in chemical plants that rely only on the faculties of science, he would probably fall short of his goal even were he to cover the entire face of the earth with factories. This is no laughing matter, for man constantly goes out of his way to commit such follies.

Man today knows that planting seeds in the ground is much easier than going through the difficult, perhaps futile, exercise of manufacturing the same seeds scientifically. He knows, but he persists in such reveries anyway.

An imitation can never outclass the original. Imperfection shall always lie in the shadow of perfection. Even though man is well aware that the human activity we call science can never be superior to nature, his attention is riveted on the imitation rather than the original because he has been led astray by his peculiar myopia that makes science appear to excel over nature in certain areas.

Man believes in the superiority of science in areas such as crop yields and aesthetics. He expects scientific farming, through the use of high-yielding techniques, to provide richer harvests than natural farming. He is convinced that taller plants can be grown by spraying hormones on rice plants grown under the forces of nature; that the number of grains per head can be increased by applying fertilizer during heading; that higher-than-natural yields can be attained by applying any of a host of yield-enhancing techniques.

Yet, no matter how many of these disparate techniques are used together, they cannot increase the total harvest of a field. This is because the amount of sunlight a field receives is fixed, and the yield of rice, which is the amount of starch produced by photosynthesis in a given area, depends on the amount of sunlight that shines on that area. No degree of human tampering with the other conditions of rice cultivation can change the upper limit in the rice yield. What man believes to be high-yielding technology is just an attempt to approach the limits of natural yields; more accurately, it is just an effort to minimize harvest losses.

So what is man likely to do? Recognizing the upper limits of yield to be set by the amount of sunlight the rice plants receive, he may well try to breach this barrier and produce yields higher than naturally possible by irradiating the rice plants with artificial light and blowing carbon dioxide over them to increase starch production. This is certainly possible in theory, but one must not forget that such artificial light and carbon gas are modeled on natural sunlight and carbon dioxide. These were created by man from other materials and did not arise spontaneously. So it is all very well and good to talk of additional increases in yield achieved over the natural limits of production by scientific technology, but because these require enormous energy outlays they are not true increases. And what is even worse, man must take full responsibility for destruction of the cyclic and material order of the natural world. This disruption in the balance of nature being the basic cause of environmental pollution, man has brought lengthy suffering down upon his own head.

The Entanglement of Natural and Scientific Farming

As I mentioned earlier, natural farming and scientific farming are diametrically opposed. Natural farming moves centripetally toward nature, and scientific farming moves centrifugally away from nature.

Yet many people think of these two approaches as being intertwined like the strands of a rope, or see scientific farming as repeatedly moving away from nature, then returning back again, something like the in-and-out motion of a piston. This is because they believe science to be intimately and inseparably allied with nature. But such thinking does not stand on a very firm foundation.

The paths of nature and of science and human action are forever parallel and never cross; and because they proceed in opposite directions, the distance between nature and science grows ever larger. As it moves along its path, science appears to maintain a cooperative association and harmony with nature, but in reality it aspires to dissect and analyze nature to know it completely in and out. Having done so, it will discard the pieces and move on without looking back. It hungers for struggle and conquest.

Thus, with every two steps forward that science takes, it moves one step back, returning to the bosom of nature and drinking of its knowledge. Once nourished, it ventures again three or four steps away from nature. When it runs into problems or out of ideas, it returns, seeking reconciliation and harmony. But it soon forgets its debt of gratitude and begins again to decry the passiveness and inefficiency of nature.

Let us take a look at an example of this pattern as seen in the development of silkworm cultivation.

Sericulture first arose when man noticed the camphor silk moth and the tussah spinning cocoons in mountain forests and learned that silk can be spun from these cocoons. The cocoons are fashioned with silk threads by moth larvae just before they enter the pupal stage. Having studied how these cocoons are made, man was no longer satisfied with just collecting natural cocoons and hit upon the idea of raising silkworms to make cocoons for him.

1. Primitive methods close to nature are believed to have marked the beginnings of sericulture. Silkworms were collected outdoors and released in woods close to home.

2. Eventually man replaced these wild species with artificially bred varieties. He noticed that silkworms thrive on mulberry leaves and that, when young, they grow more rapidly if these leaves are fed to them finely chopped. At this point, it became easier to raise them indoors, so man built shelves that allowed him to grow large numbers of worms inside. He devised feeding shelves and special tools for cocoon production, and became very concerned about observing optimum temperature and humidity. The methods used during this long period of sericulture development demanded a great deal of hard labor from farming households. One had to get up very early in the morning, shoulder a large mulberry basket, and walk out to the mulberry grove, there to pick the leaves one at a time. The leaves were carefully wiped of dew with dry cloths, chopped into strips with a large knife, and scattered over the silkworms on the tens and hundreds of feeding shelves.

The grower carefully maintained optimum conditions night and day, taking the greatest pains to adjust room temperature and ventilation by installing heaters and opening and closing doors. He had no choice; the silkworms improved by artificial breeding were weak and susceptible to disease. Many were the times when, after having finally grown to full size, the silkworms were suddenly wiped out by disease. During spinning of the silk from the cocoons, all the members of the family pitched in, rarely getting any sleep. Growing and care of the mulberry trees also kept farmers busy with fertilizing and weeding. If a late frost killed the young leaves, then one usually had no choice but to throw away the whole lot of silkworms.

Given such labor-intensive methods, it should come as no surprise then that people began to look for less strenuous techniques. Over the last 15 to 20 years, sericulture techniques that approach natural farming have spread widely among growers.

3. These methods consist of, for example, throwing branches of mulberry leaves onto the silkworms rather than picking and chopping leaves. Once it was learned that such a crude method works for young silkworms as well as the fully grown larvae, the next thought that occurred to growers was that instead of raising silkworms in a special room, they might perhaps be raised outdoors in a small shed, under the eaves, or in a sort of hotbed. On trying the idea out, farmers found that silkworms are really quite hardy and there never was a need in the first place to raise them under constant temperature and humidity conditions. Needless to say, growers were overjoyed. Originally a creature of nature, the silkworms thrived outdoors day and night; only man feared the evening dew.

As advances were made in rearing methods, silkworms were raised first under the eaves, then outdoors, and finally were released into nearby trees. Sericulture appeared to be heading in the direction of natural farming, when all of a sudden the industry fell upon hard times. The rapid development of synthetic fibers almost made natural silk obsolete. The price of silk plummeted, throwing sericulture farms out of business. Raising silkworms became regarded as something of a backwards industry.

However, the growing material affluence of our times has nurtured extravagant tastes in people. Consumers rediscovered the virtues of natural silk absent in synthetic fibers, causing silk to be treated once again as something of a precious commodity. The price of silk cocoons skyrocketed and people regained an interest in silkworms.

4. Yet by this time the hard-working farmer of old was gone, so innovative new sericulture techniques were adopted. These are purely scientific methods that go in a direction opposite to that of natural farming: industrial sericulture. Artificial feed is prepared from mulberry leaf powder, soybean powder, wheat powder, starch, fats, vitamins, and other ingredients. It also contains preservatives and is sterilized. Naturally, the silkworms are raised in a plant fully outfitted with heating and air conditioning equipment, and lighting and ventilation are adjusted automatically. Feed is carried in, and droppings carried out, on a belt conveyor.

Fig. 3.3   Natural farming moves inward towards "nothingness" and scientific farming moves outward towards infinity.

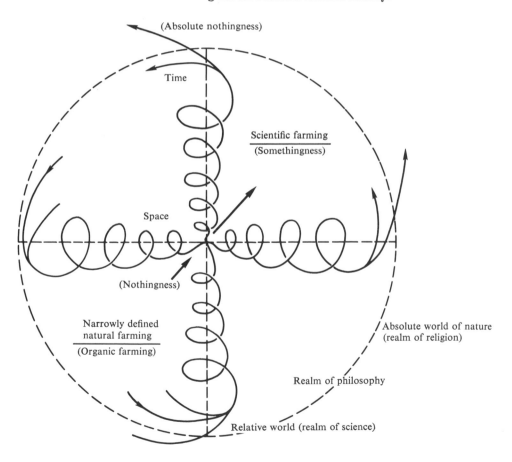

If a disease should break out among the silkworms, the room can be hermetically sealed and disinfected with gas. With all feeding and cocoon collection operations totally automated, we have reached an age in which natural silk is something produced in factories. Although the starting material is still mulberry leaves, this will probably be replaced by a totally synthetic feed prepared from petrochemicals. Once an inexhaustible supply of cocoons can be produced in factories from a perfect diet, human labor will no longer be required. Will people then rejoice at how easily and effortlessly silk can be had in any amount?

Sericulture has in this way shifted repeatedly from one course to another. From natural farming it moved to scientific farming, then appeared to move a step back in the direction of natural farming. However, once scientific farming begins to get under way, it does not regress or turn back, but rushes madly along a path that takes it away from nature.

The intertwining of natural farming and scientific farming can be depicted as shown in Fig. 3.3. Narrowly defined natural farming, which includes organic farming, proceeds centripetally inward towards a state of "nothingness" by the elimination of human labor; it compresses and freezes time and space. Modern scientific farming, on the other hand, seeks to appropriate time and space through complex and diverse means; it proceeds centrifugally outward towards "something-ness," expanding and developing as it goes. Both can be understood as existing in a relative relationship in the same dimension or plane. But although the two may appear identical at a given point, they move in opposite directions, the one headed for zero and the other for infinity.

Thus, seen relatively and discriminatively, the two readily appear to be in opposition, yet intimately intertwined, neither approaching nor moving away from one another, advancing together and complementarily through time. However, because natural farming condenses inward, seeking ultimately a return to the true world of nature that transcends the world of relativity, it is in irreconcilable conflict with scientific farming, which expands forever in the relative world.

2. The Four Principles of Natural Farming

I have already shown how natural farming is clearly and undeniably superior to scientific farming, both in theory and in practice. And I have shown that scientific farming requires human labor and large expenditures, compounds chaos and confusion, and leads eventually to destruction.

Yet man is a strange creature. He creates one troublesome condition after another and wears himself down observing each. But take all these artificial conditions away and he suddenly becomes very uneasy. Even though he may agree that the natural way of farming is legitimate, he seems to think that it takes extraordinary resolve to exercise the principle of "doing nothing."

It is to allay this feeling of unease that I recount my own experiences. Today, my method of natural farming has approached the point of "doing nothing." I will admit that I have had my share of failures during the forty years that I have been at it. But because I was headed in basically the right direction, I now have yields that are at least equal to or better than those of crops grown scientifically in every respect. And most importantly:

1) My method succeeds at only a tiny fraction of the labor and costs of scientific farming, and my goal is to bring this down to zero.

2) At no point in the process of cultivation or in my crops is there any element that generates the slightest pollution, in addition to which my soil remains eternally fertile.

There can be no mistaking these results, as I have achieved them now for a good many years. And I guarantee that anyone can farm this way. This method of "do-nothing" farming is based on four major principles:
1. No cultivation
2. No fertilizer
3. No weeding
4. No pesticides

No Cultivation

Plowing a field is hard work for the farmer and usually one of the most important activities in farming operations. In fact, to many people, being a farmer is synonymous with turning the soil with plow or hoe. If working the soil is unnecessary then, the image and reality of the farmer change drastically. Let us look at why plowing is thought to be essential and what effect it actually has.

Plowing Ruins the Soil: Knowing that the roots of crops penetrate deep into the earth in search of air, water, and nutrients, people reason that making larger amounts of these ingredients available to the plants will speed crop growth. So they clear the field of weeds and turn the soil from time to time, believing this loosens and aerates the soil, increases the amount of available nitrogen by encouraging nitrification, and introduces fertilizer into the soil where it can be absorbed by the crops.

Of course, plowing under chemical fertilizers scattered over the surface of a field will probably increase fertilizer effectiveness. But this is true only for cleanly plowed and weeded fields on which fertilizer is applied. Grassed fields and no-fertilizer cultivation are a different matter altogether. We therefore have to examine the necessity of plowing from a different perspective. As for the increase in available nitrogen through nitrification, this is analogous to wasting one's body for some temporary gain.

Plowing is supposed to loosen the soil and improve the penetration of air, but does not this in fact have the opposite effect of compacting the soil and decreasing air porosity? When a farmer plows his fields and turns the soil with a hoe, this appears to create air spaces in the soil and soften the dirt. But the effect is the same as kneading bread: as the farmer turns the soil with his hoe, this breaks the soil up into smaller and smaller particles which acquire an increasingly regular physical arrangement with smaller interstitial spaces. The result is a harder, denser soil.

The only effective way to soften up the soil is to apply compost and work it into the ground by plowing. But this is only a short-lived measure. In fields that have been weeded clean and carefully plowed and re-plowed, the natural aggregation of the soil into larger particles is disturbed, and the soil particles become finer and finer, hardening the ground.

Wet paddy fields are normally supposed to be tilled five, six, or even seven times during the growing season. The more zealous farmers even compete with each other to increase the number of plowings. Everyone thought this softened the soil in the paddy and let more air into the soil. And that is the way it looked to most people for a long time, until after World War II, when herbicides became available. Then farmers discovered that when they sprayed their fields with herbicides, the less they plowed, the better were their yields. This just went to show that intertillage had been effective as a weeding process but had been worthless as a means for loosening the soil.

To say that tilling the soil is worthless is not the same as saying that it is unnecessary to loosen the soil and increase its porosity. No, in fact I would like to stress, more than anyone else, just how important an abundance of air and water are to the soil. It is in the nature of soil to swell and grow more porous with each passing year. This is absolutely essential for microorganisms to multiply in the earth, for the soil to grow more fertile, and for the roots of large trees to penetrate deep into the ground. Only, I believe that, far from being the answer, working the soil with plow and hoe actually interferes with these processes. If man lets the soil take care of itself, it will enrich and loosen itself with the powers of nature.

Farmers usually plow the soil to a depth of about four to eight inches, whereas the roots of grasses and green manure crops work the soil down to twelve inches, fifteen inches, or more. When these roots reach down deep into the earth, air and water penetrate into the soil together with the roots. As these wither and die, many types of microorganisms proliferate. These die off and are replaced by others, increasing the amount of humus and softening the soil. Earthworms eventually appear where there is humus, and as the number of earthworms increases, moles begin burrowing through the soil.

The Soil Works Itself: The soil lives of its own accord and plows itself. It needs no help from man. Farmers often talk of "taming the soil" and of a field becoming "mature," but why is it that trees in mountain forests grow to such magnificent heights without the benefit of hoe or fertilizer, while the farmer's fields can grow only puny crops?

Has the farmer ever given any careful thought to what plowing is? Has not he trained all his attention on a thin surface layer and neglected to consider what lies below that?

Trees seem to grow almost haphazardly in the mountains and forests, but the cedar grows where it can thrive to its great size, clumps of mixed trees rise up where they were meant to grow, and pine trees germinate and grow in places suited for pine trees. One does not see pines growing in valley bottoms or cedar seedlings taking root on mountain tops. One type of fern grows on infertile land and another in areas of deep soil. Plants that normally grow along the water's edge do not grow on mountain tops, and terrestrial plants do not grow in the water. Although apparently without intent or purpose, these plants know exactly where they can and should grow.

Man talks of "the right crop for the right land," and does studies to determine which crops grow well where. Yet research has hardly touched upon such topics as the type of parent rock and soil structure suited to mandarin orange trees, or the physical, chemical, and biological soil structures in which persimmon trees grow well. People plant trees and sow seed without having the faintest idea of what the parent rock in their field is and without knowing anything about the structure of the soil. It is no wonder then that farmers worry about how their crops are going to turn out.

In the mountain forests, however, concerns over the physical and chemical compositions of the topsoil and deeper strata are nonexistent; without the least help from man, nature creates the soil conditions sufficient to support dense stands of towering trees. In nature, the very grasses and trees, and the earthworms and moles in the ground, have acted the part of plowhorse and oxen, completely rearranging and renewing the soil. What can be more desirable to the farmer than being able to work the fields without pulling a plow or swinging a hoe? Let the grasses plow the topsoil and the trees work the deeper layers. Everywhere I look, I am reminded of how much wiser it is to entrust improvement of the soil to the soil and plant growth to the inherent powers of plants.

People transplant saplings without giving a thought as to what they are doing. They join a graft to the stock of another species or clip the roots of a fruit sapling and transplant it. From this point on, the roots cease to grow straight and lose the ability to penetrate hard rock. During transplanting, even a slight entanglement of a tree's roots interferes with the normal growth of the first generation of roots and weakens its ability to send roots deep into the soil. Applying chemical fertilizers encourages the tree to grow a shallow root structure that extends along the topsoil. Fertilizer application and weeding bring a halt to the normal aggregation and enrichment of topsoil. Clearing new land for agriculture by pulling up trees and bushes robs the deeper layers of the soil of a source of humus, halting the active proliferation of soil microbes. These very actions are what make plowing and turning the soil necessary in the first place.

There is no need to plow or improve a soil because nature has been working at it with its own methods for thousands of years. Man has restrained the hand of nature, and taken up the plow himself. But this is just man imitating nature. All he has really gained from all this is a mastery at scientific exposition.

No amount of research can teach man everything there is to know about the soil, and he will certainly never create soils more perfect than those of nature. Because nature itself is perfect. If anything, advances in scientific research teach man just how perfect and complete a handful of soil is, and how incomplete human knowledge is.

We can either choose to see the soil as imperfect and take the hoe in hand, or trust the soil and leave the business of working it to nature.

No Fertilizer

Crops Depend on the Soil: When we look directly at how and why crops grow on the earth, we realize that they do so independently of human knowledge and action. This means that they have no need basically for such things as fertilizers and nutrients. Crops depend on the soil for growth.

I have experimented with fruit trees and with rice and winter grain to determine whether these can be cultivated without fertilizers.

Of course crops can be grown without fertilizer. Nor does this yield the poor harvests people generally believe. In fact, I have been able to show that by taking full advantage of the inherent powers of nature, one can obtain yields equal to those that can be had with heavy fertilization. But before getting into a discussion of why it is possible to farm without using fertilizers, and whether the results are good or bad, I would like to look first at the course scientific farming has taken.

People first saw crops growing in the wild and called this "growth." Applying discriminating knowledge, they proceeded from the notion of wild plant growth to plant cultivation.

For example, scientists will typically begin by analyzing rice and barley plants and identifying the various nutrients. They will then speculate that these nutrients promote the growth of rice and barley. Next they will apply the nutrients as fertilizer, and observing that the rice or barley grows as they had expected, will conclude that the fertilizer is what makes the crops grow. From the moment they compare crops grown with and without fertilizer and conclude that fertilizer application results in taller, better yielding plants than no application, people cease to doubt the value of fertilizers.

Are Fertilizers Really Necessary?:   The same is true when one delves into the origins of why fertilizers are thought to be essential to fruit trees. Pomologists normally begin with an analysis of the trunk, leaves, and fruit of the tree. From this they learn what the nitrogen, phosphorus, and potassium contents are and how much of these components are consumed per unit of annual growth or of fruit produced. Based on the results of such analyses, fertilization plans for fruit trees in mature orchards will typically set the amount of nitrogen components at 90 pounds, say, and the amount of phosphates and potassium at 70 pounds each. Researchers will apply fertilizer to trees grown in test plots or earthen pots, and examining the growth of the tree and the amount and quality of fruit it bears, will claim to have demonstrated the indispensability of fertilizer.

Learning that nitrogenous components are present in the leaves and branches of citrus trees and that these are absorbed from the ground by the roots, man hits upon the idea of administering fertilizer as a nutrient source. If this succeeds in supplying the nutrient needs of the leaves and branches, man immediately jumps to the conclusion that applying fertilizer to citrus trees is both necessary and effective.

If one works from the assumption that fruit trees "must be grown," the absorption of fertilizer by the roots becomes the cause, and the full growth of the leaves and branches the effect. This leads quite naturally to the conclusion that applying fertilizer is necessary.

However, if we take as our starting point the view that a tree grows of its own accord, the uptake of nutrients by the tree's roots is no longer a cause but, in the eyes of nature, just a small effect. One could say that the tree grew as a result of the absorption of nutrients by the roots, but one could also claim that the absorption of nutrients was caused by something else, which had the effect of making the tree grow. The buds on a tree are made for budding and so this is what they do; the roots, with their powers of elongation, spread and extend throughout the earth. A tree has a shape perfectly adapted to the natural environment. With this, it guards the providence of nature and obeys nature's laws, growing neither too fast nor too slow, but in total harmony with the great cycles of nature.

The Countless Evils of Fertilizer: What happens when the farmer arrives in the middle of all this and spreads his fields and orchards with fertilizer? Dazzled and led astray by the rapid growth he hears of, he applies fertilizer to his trees without giving any thought to the influence this has on the natural order.

As long as he cannot know what effects scattering a handful of fertilizer has on the natural world, man is not qualified to speak of the effectiveness of fertilizer application. Determining whether fertilizer does a tree or soil good or harm is not something that can be decided overnight.

The more scientists learn, the more they realize just how awesome is the complexity and mystery of nature. They find this to be a world filled with boundless, inscrutable riddles. The amount of research material that lies hidden in a single gram of soil, a single particle, is mind-boggling.

People call the soil mineral matter, but some one hundred million molds, bacteria, yeasts, diatoms, and other microbes live in just one gram of ordinary topsoil. Far from being dead and inanimate, soil is teeming with life. These microorganisms do not exist without reason. Each lives for a purpose, struggling, cooperating, and carrying on the cycles of nature.

Into this soil, man throws powerful chemical fertilizers. It would take many years of research to determine how the fertilizer components combine and react with air, water, and many other substances in nonliving mineral matter, what changes they undergo, and what relationships should be maintained between these components and the various microorganisms in order to guard a harmonious balance.

Very little, if any, research has been done yet on the relationship between fertilizers and soil microbes. In fact, most experiments totally ignore this. At agricultural research stations, scientists place some soil in pots and run some tests, but more likely as not, most of the soil microbes in these pots die off. Clearly, results obtained from tests conducted under fixed conditions and within a limited framework cannot be applied to situations under natural conditions.

Yet, just because a fertilizer slightly accelerates crop growth in such tests, it is praised lavishly and widely reported to be effective. Only the efficacy of the fertilizer is stressed; almost nothing is said about its adverse effects, which are innumerable. Here is just a sampling:

1.   Fertilizers speed up the growth of crops, but this is only a temporary and local effect that does not offset the inevitable weakening of the crops. This is similar to the rapid acceleration of plant growth by hormones.

2.   Plants weakened by fertilizers have a lowered resistance to diseases and insect predators, and are less able to overcome other obstacles to growth and development.

3.   Fertilizer applied to soil usually is not as effective as in laboratory experiments. For example, it was recently learned that some thirty percent of the nitrogenous component of ammonium sulfate applied to paddy fields is denitrified by microorganisms in the soil and escapes into the atmosphere. That this came out after decades of use is an unspeakable injury and injustice to countless farmers that cannot be laughed off as just an innocent mistake. Such nonsense will occur again and again. Recent reports say that phosphate fertilizers applied to fields only penetrate two inches into the soil surface. So it turns out that all those mountains of phosphates that farmers religiously spread on their fields year after year are useless and are essentially being "dumped" on the topsoil.

4.   Damage caused directly by fertilizers is also enormous. More than seventy percent of the "big three"—ammonium sulfate, superphosphate, and potassium sulfate—is concentrated sulfuric acid which acidifies and causes great harm to the soil, both directly and indirectly. Each year, some 1.8 million tons of sulfuric acid are dumped onto the farmlands of Japan in the form of fertilizer. This acidic fertilizer suppresses and kills soil microorganisms, disrupting and damaging the soil in a way that may one day spell disaster for Japanese agriculture.

5.   One major problem with fertilizer use is the deficiency of trace components. Not only have we relied too heavily on chemical fertilizers, killing the soil, our production of crops from a small number of nutrients has led to a deficiency in many trace elements essential to crops. Recently, this problem has risen to alarming proportions in fruit trees, and has also surfaced in rice cultivation as one cause of low harvests.

The effects and interactions of the various components of fertilizers in orchard soil are unspeakably complex. Nitrogen and phosphate uptake is poor in iodine-deficient soils. When the soil is acidic or turns alkaline through heavy applications of lime, deficiencies of zinc, manganese, boron, iodine, and other elements develop because these become less soluble in water. Too much potassium blocks iodine uptake and reduces the absorption of boron as well. The greater the amount of nitrogen, phosphates, and potassium administered to the soil, the higher the resulting deficiency of zinc and boron. On the other hand, higher levels of nitrogen and phosphates result in a lower manganese deficiency.

Adding too much of one fertilizer renders another fertilizer ineffective. When there is a shortage of certain components, it does no good to add a generous amount of other components. As scientists get around to studying these relationships, they begin to realize just how complex the addition of fertilizers is. If we were prudent enought to apply fertilizers only when we were certain of the pros and cons, we could be sure of avoiding dangerous mistakes, but the benefits and dangers of fertilization are never likely to become perfectly clear.

And the problems go on multiplying. Very limited research is currently underway on several trace components, but an endless number of such components remain to be discovered. This will spawn infinite new areas of study, such as mutual interactions, leaching in the soil, fixation, and relationships with microbes. Still, in spite of such intimidating complexity, if a fertilizer happens to be effective in one narrowly designed experiment, scientists will report this to be remarkably effective without having the vaguest idea of its true merits and weaknesses.

"Well yes," the farmer all too easily reasons, "chemical fertilizers do cause some damage, but I've used fertilizers now for years and haven't had any big problems, so I suppose that I'm better off with them." The seeds of calamity have been sown and are sending up shoots that are about to emerge. When we take note of the danger, it will be too late to turn the tables on misfortune.

On top of this there is the fact that farmers have always had to struggle to scrape together enough to buy fertilizer. Why, to give one simple example, fertilizers currently account for thirty to fifty percent of the costs of running an orchard.

People claim that produce cannot be grown without fertilization, but is it really true that crops do not grow in the absence of fertilizer? Is the use of fertilizers economically advantageous? And have methods of farming with fertilizers made the lot of farmers easier?

Why the Absence of No-Fertilizer Tests?:   Strange as it may seem, scientists hardly ever run experiments on no-fertilizer cultivation. In Japan, only a handful of reports have been published over the last few years on the cultivation of fruit trees without fertilizer in small concrete enclosures and earthen pots. Some tests have been done on rice and other grains, but only as controls. Actually, the reason why no-fertilizer tests are not performed is all too clear. Scientists work from the basic premise that crops are grown with fertilizer. Why experiment with such an idiotic and dangerous method of cultivation, they say. Why indeed.

The standard on which fertilizer experiments should be based is no-fertilizer tests, but three-element tests are the standard actually used.

Quoting the results of a very small number of insignificant experiments, people claim that a tree grows only about half as much without fertilizer as when various types of fertilizer are used, and the common belief is that yields are terrible—on the order of one third that obtained with fertilizers. However, the conditions under which these no-fertilizer experiments were conducted have little in common with true natural farming.

When crops are planted in small earthenware pots or artificial enclosures, the soil in which they grow is dead soil. The growth of trees whose roots are boxed in by a concrete enclosure is highly unnatural. It is unreasonable to claim that because plants grown without fertilizer in such an enclosure grow poorly, they cannot be grown without fertilizers.

No-fertilizer natural farming essentially means the natural cultivation of crops without fertilizers in a soil and environment under totally natural conditions. By totally natural cultivation I mean no-fertilizer tests under condition-less conditions. However such experiments are out of the reach of scientists, and indeed impossible to perform.

I am convinced that cultivation without fertilizers under natural circumstances is not only philosophically feasible, but is more beneficial than scientific, fertilizer-based agriculture, and is preferable for the farmer. Yet, although cultivation without the use of chemical fertilizers is possible, crops cannot immediately be grown successfully without fertilizers on fields that are normally plowed and weeded.

It is imperative that farmers think seriously about what nature is and provide a growing environment that approaches at least one step closer to nature. But to farm in nature, one must first make an effort to return to that natural state which preceded the development of the methods of farming used by man.

Take a Good Look at Nature:　When trying to prove whether crops can be grown without fertilizers, one cannot tell anything by looking only at the crops. One must begin by taking a good look at nature.

The trees of the mountain forests grow under conditions close to pure nature, receiving no fertilizer by the hand of man. Yet they grow very well year after year. Reforested cedars in a favorable area generally grow about forty tons per quarter acre over a period of twenty years. These trees thus produce about two tons of growth each year without fertilizer. This includes only that part of the tree that can be used as lumber, so if we also take into account small branches, leaves, and roots, then annual production is probably close to double, or about four tons.

If we were talking of a fruit orchard here, then this would translate into two to four tons of fruit produced each year without fertilizers—about equal to standard production levels by fruit growers today.

After a certain period of time, the wood in a timber stand is felled, and the entire surface portion of the tree—including the branches, leaves, and trunk—is carried away. So not only are fertilizers not used, this is slash-and-burn agriculture. How then, and from where, are the fertilizer components for this production volume supplied each year to the growing trees? Plants do not need to be raised, they grow of their own accord. The mountain forests are living proof that trees are not raised with fertilizer but grow by themselves.

One might also point out that because the planted cedars are not virgin forest, they are not likely to be growing under the full powers of the natural soil and environment. The damage caused by repeated planting of the same species of tree, the felling and harvesting of the timber, and the burning of the mountainside take their toll. Anyone who sees Morishima acacia planted in depleted soil on a mountainside and succeeded a number of years later with giant cedars many times their size will be amazed at the great productive powers of the soil. When acacia is planted among cedar or cypress, these latter thrive with the help of microbes on the roots of the acacia. If the mountain forest is left to itself, the action of the wind and snow over the years weathers the rocks, a layer of humus forms and deepens with the fall of leaves each year, microorganisms multiply in the soil—turning it a rich black, and the soil aggregates and softens, increasing water retention. There is no need for human intervention here, and the trees grow on and on.

Nature is not dead. It lives and it grows. All that man has to do is direct these vast hidden forces to the growth of fruit trees. But rather than using this great power, people choose to destroy it. Weeding and plowing the fields each year depletes the fertility of the soil, creates a deficiency of trace components, diminishes the soil's vitality, hardens the topsoil, kills off microbes, and turns rich, living, organic material into a dead, inanimate, yellowish-white mineral matter the only function of which is to physically support the crops.

Fertilizer Was Never Necessary to Begin With: Let us take a look at the farmer as he clears a forest and plants fruit trees. He fells the trees in the forest and carries them off as logs, taking the branches and leaves as well. Then he digs deep into the earth, pulling up the roots of trees and grasses, which he burns. Next, he turns the soil over and over again to loosen it up. But in so doing, he destroys the physical structure of the soil. After pounding and kneading the soil again and again like bread dough, he drives out air and the humus so essential to microorganisms, reducing it to a yellow mineral matter barren of life. He then plants fruit saplings in the now dead soil, adds fertilizer, and attempts to grow fruit trees entirely through his own efforts.

At agricultural test centers, fertilizer is added to potted soil after that soil has become mineral matter devoid of life and nutrients. The effort is like sprinkling water on dry soil: the trees thrive on the fertilizer nutrients. Naturally, the researchers report this as evidence of the remarkable effectiveness of the fertilizer. The farmer simulates the laboratory procedure by carefully clearing the land of all plant matter and killing the soil in the field, then applying fertilizer. Naturally, he notes the same startling results and is pleased with what he sees.

The poor farmer has taken the long way around. Although I would not call fertilizers totally useless, the fact is that nature provides us with all the fertilizers we need. Crops grow very well without chemical fertilizers. Since ancient times, rock outcroppings on the earth have been battered by winds and rain, first into boulders and stones, then into sand and earth. As this gave rise to and nurtured microbes, grasses, and eventually great, towering trees, the land became buried under a mantle of rich soil.

Even though it is unclear how, when, and from where the nutrients essential to plant growth are formed and accumulate, each year the topsoil becomes darker and richer. Compare this with the soil in the fields farmed by man, which grows poorer and more barren each year, in spite of the large amounts of fertilizer constantly poured on.

The no-fertilizer principle does not say that fertilizers are worthless, but that there is no need to apply chemical fertilizers. Scientific technology for applying fertilizers is basically pointless for the same reason. Yet research on the preparation and use of organic composts, which are much closer to nature, appears at first glance to be of value.

When compost such as straw, grasses and trees, or seaweed is applied directly to a field, this takes a while to decompose and trigger a fertilizer response in the crops. This is because microorganisms help themselves to the available nitrogen in the soil, creating a temporary nitrogen deficiency that initially starves the crops of needed nitrogen. In organic farming, these materials are therefore fermented and used as prepared compost, giving a safe, effective fertilizer.

All the trouble taken during preparation of the compost to speed up the rate of fertilizer response, such as frequent turning of the pile, methods for stimulating the growth of aerobic bacteria, the addition of water and nitrogenous fertilizers, lime, superphosphate, rice bran, manure, and so forth—all this trouble is taken just for a slight acceleration in response. Since the net effect of these efforts is to speed up decomposition by at most ten to twenty percent, this can hardly be called necessary. Especially since there already was a method of applying straw to the fields that achieved outstanding results.

The logic that rejects grassed fields, green manure, and the direct application and plowing under of human wastes and livestock manure changes with the circumstances and times. Given the right conditions, these may be effective, but no fertilizer method is absolute. The surest way to solve the problem is to apply a method that adapts to the circumstances and follows nature.

I firmly believe that, while compost itself is not without value, the composting of organic materials is fundamentally useless.

No Weeding

Nothing would be more welcome to the farmer than not having to weed his fields, for this is his greatest source of toil. Not having to weed or plow might sound like asking for too much, but if one stops to think about what repeatedly weeding and running a plow through a field actually means, it becomes clear that weeding is not as indispensable as we have been led to believe.

Is There Such a Thing as a Weed?:   Does no one question the common view that weeds are a nuisance and harmful to the raising of crops?

Man distinguishes between crops and weeds, and the first step he takes in that respect is to decide whether to weed or not to weed. Like the many different microorganisms that struggle and cooperate in the soil, myriad grasses and trees live together on the soil surface. Is it right then to destroy this natural state, to pick out certain plants living in harmony among many plants, to call these "crops" and uproot all the others as "weeds"?

In nature, plants live and thrive together, but man sees things differently. He sees coexistence as competition; he thinks of one plant as hindering the growth of another and believes that to raise a crop, he must remove other grasses and herbs. Had man looked squarely at nature and placed his trust in its powers, would he not have raised crops in harmony with other plants? Ever since he chose to differentiate crop plants from other plants, he has felt compelled to raise crops through his own efforts. When man decides to raise one crop, the attention and devotion he focuses on raising that crop gives birth to a complementary sense of repulsion and hate that excludes all else.

The moment that the farmer started caring for and raising his crops, he began to regard other herbs with disgust as weeds and has strived ever since to remove them. But the growth of weeds being natural, there is no end to their variety or to the labors of those who work to remove them.

If one believes that crops grow with the aid of fertilizers, then the surrounding weeds must be removed because they rob the crop plants of fertilizer. But in natural farming, where plants grow of their own accord without relying on fertilizers, the surrounding weeds do not pose any problem at all. Nothing is more natural than to see grass growing at the foot of a tree; no one would even think of that grass as interfering with the growth of the tree.

Actually, in nature, bushes and shrubs grow at the foot of large trees, grasses spread among the shrubs, and mosses fluorish beneath the grasses. Instead of cut-throat competition for nutrients, this is a peaceful scene of coexistence. Rather than seeing the grasses as stunting shrub growth, and the shrubs as slowing the growth of trees, one should feel instead a sense of wonder and amazement at the ability of these plants to grow together in this way.

Grasses Enrich the Soil: Rather than pulling weeds, people should give some thought to the significance of these plants. Having done so, they will agree that the farmer should let the weeds live and make use of their strength. Although I call this the "no-weeding" principle, it could also be known as the principle of "weed utility."

Long ago, when the earth began to cool and the surface of the earth's crust weathered, forming soil, the first forms of life to appear were the bacteria and unicellular plants such as diatoms. All plants arose for a reason, and all plants live and thrive today for a reason. None is useless; each makes its own contribution to the development and enrichment of the biosphere. Such fertile soil would not have formed on the earth's surface had there been no microorganisms in the earth and grasses above it. Grasses and other plants do not grow without a purpose.

The deep penetration of grass roots into the earth loosens the soil. When the roots die, this adds to the humus, allowing soil microbes to proliferate and enrich the soil. Rainwater percolates through the soil and air is carried deep down, supporting earthworms, which eventually attract moles. Weeds are absolutely essential for a soil to be organic and alive.

Without grasses growing over the surface of the ground, rainwater would wash away a share of the topsoil each year. Even in gently sloping areas, this would result in the loss of from several tons to perhaps well over a hundred tons per year. After twenty or thirty years, the topsoil washes entirely away, reducing soil fertility to essentially zero. Essential as they are then, it would make more sense for farmers to stop pulling weeds and begin making use of their considerable powers.

It is understandable, however, when farmers say that weeds growing wild in rice and wheat fields or under fruit trees interfere with other work. Even where cultivation with weeds appears to be possible and perhaps even beneficial in principle, monoculture is more convenient for the farmer. This is why, in practice, one must adopt a method that makes use of the strength of weeds but also takes into account the convenience of farming operations: a "weedless" method that allows the weeds to grow.

A Cover of Grass Is Beneficial: This method includes sod and green manure cultivation. In my citrus orchard, I attempted first cultivation under a cover of grass, then switched to green manure cultivation, and now I use a ground cover of clover and vegetables with no weeding, tillage, or fertilizer. When weeds are a problem, then it is wiser to remove weeds with weeds than to pull weeds by hand.

The many different grasses and herbs in a natural meadow appear to grow and die in total confusion, but upon closer examination, there are laws and there is order here. Grasses meant to sprout do so, those that fluorish do so for a reason, and if plants weaken and die, there is a cause. Plants of the same species do not all grow in the same place and way, but given types fluorish, then fade in an ongoing succession. The cycles of coexistence, competition, and mutual benefit repeat themselves. Certain weeds grow as individuals, others grow in bunches, and yet others form colonies. Some grow sparsely, some densely, and some in clumps. Each has a different ecology: some grow over their neighbors and over-power them, some wrap around others in symbiosis, some weaken other plants, and some die—while others thrive—as undergrowth.

By studying and making use of the properties of weeds, one weed can be used to drive out a large number of other weeds. If the farmer were to grow grasses or green manure crops that take the place of undesirable weeds and are beneficial to him and his crops, then he would no longer have to weed, in addition to which the green manure would enrich the soil and prevent its erosion. I have found that by "killing two birds with one stone" in this way, growing fruit trees and tending an orchard can be made easier and more beneficial than normal methods. In fact, from my experience, there is no question that weeding in orchards is not only useless, it is positively harmful.

What about in the case of crops such as rice or barley? I believe that the coex-istence of surface plants is true to nature, and that the no-weeding principle applies also to rice and barley cultivation. But because the presence of weeds among the rice and barley interferes with harvesting, these weeds have to be re-placed with some other herb.

I practice a form of rice-barley succession cropping in which I seed barley to-gether with clover over the standing heads of rice, and scatter rice seed and green manure while the barley is up. This more nearly approaches nature and eliminates weeding.

My reason for trying such a method was not that I was tired of weeding or wanted to prove that cultivation is possible without weeding. I did this out of dedication to my goals of understanding the true form of rice and barley and of achieving more vigorous growth and higher yields by cultivating these grains in as natural a way as possible.

What I found was that, like fruit trees, rice and barley too can be grown with-out weeding. I learned also that vegetables can be grown in a state that allows them to go wild, without fertilizer or weeding, and yet yields comparable to normal methods attained.

No Pesticides

Insect Pests Do Not Exist: The moment the problem of crop disease or insect damage arises, talk turns immediately to methods of control. But we should begin by examining whether crop disease or insect damage exist in the first place. A thousand plant diseases exist in nature, but in truth there are none. It is the agricultural specialist who gets carried away with discussions on disease and pest damage. Although research is done on ways to reduce the number of country villages without doctors, no studies are ever run to determine how these villages have gotten along without doctors. In the same way, when people spot signs of a plant disease or an insect pest, they immediately go about trying to get rid of it. The smart thing to do would be to stop treating insects as pests and find a way that eliminates the need for control measures altogether.

I would like to take a look now at the issue of new pesticides, which has escalated into a major pollution problem. The problem exists because, very bluntly, there are no non-polluting new pesticides.

Most people seem to believe that the use of natural enemies and pesticides of low toxicity will clear up the problem, but they are mistaken. Many feel reassured by the thought that the use of beneficial insect predators to control pests is a biological method of control without harmful repercussions, but to someone who understands the chain of being that links together the world of living organisms, there is no way of telling which organisms are natural enemies and which are pests. By meddling with controls, all man accomplishes is destruction of the natural order. Although he may appear to be protecting the natural enemies and killing the pests, there is no knowing whether the pests will become beneficial and the predators pests. Many insects that are harmless in a direct sense are harmful indirectly. And when things get even more complex, as when one beneficial insect feeds on a pest that kills another beneficial insect which feeds on another pest, it is futile to try and draw sharp distinctions between these and apply pesticides selectively.

Pollution by New Pesticides: With the problem of pesticide pollution, many await the development of new pesticides that:

1. have no adverse effects on animal cells and act by inhibiting enzymes specific to given insects, microorganisms, pathogens, plants, or whatever;

2. are degradable under the action of sunlight and microorganisms, and are totally non-polluting, leaving no residues.

The antibiotics plastocysin and kasgamycin were released onto the market as new pesticides that meet these conditions, and used widely as a preventive measure against rice blast disease amid great clamor and publicity. Another recent area of investigation in which many are placing much hope is pesticides prepared from biological components already present in nature, such as amino acids, fatty acids, and nucleic acids. Such pesticides, it is generally surmised, are not likely to leave residues.

One other new type of pesticide discovered recently and reported as possibly non-polluting is a chemical that suppresses metamorphosis-regulating hormones in insects. Insects secrete hormones that control the various stages of metamorphosis, from the egg to the larva, the pupa, and finally the adult. A substance extracted from the bay tree apparently inhibits secretion of these hormones.

Because these substances work selectively on only certain types of insects, they are thought to have no effects on other animals and plants. But this is incorrect and short-sighted. Animal cells, plant cells, and microorganisms are basically all quite similar. When a pesticide that works on some insect or pathogen is said to be harmless to plants and animals, this is merely a word game that plays on a very minor difference in resistance to that substance.

A substance that is effective on insects and microorganisms also acts, to a greater or lesser degree, on plants and animals. A pesticidal or bactericidal effect is referred to as phytotoxicity in plants and pollution in animals and man.

It is unreasonable to expect a substance to work only on specific insects and microbes. To claim that something does not cause pesticide damage or pollution is to make small distinctions based on minor differences in action. Moreover, there is no knowing when these minor differences will change or turn against us. Yet, in spite of this constant danger, people are satisfied if a substance poses no immediate threat of damage or pollution and do not bother to consider the greater repercussions of its effects. This attitude of ready acceptance complicates the problem and aggravates the dangers.

The same is true as well of microorganisms employed as biological pesticides. Many different types of bacteria, viruses, and molds are sold and used in a variety of applications, but what effect are these having on the biosphere? One hears a lot lately about pheromones. These are chemicals produced by organisms in minute quantities that trigger very profound physiological changes or specific behavioral reactions in other individuals. They may be used, for example, to attract the males or females of a given insect pest. Even the use of chemosterilants together with such attractants and excitants is conceivable.

Sterilization can be achieved by a number of methods, such as destruction of the reproductive function by irradiation with gamma rays, the use of chemosterilants, and interspecific mating. But no evidence exists to support the claim that the effects of sterilization are limited to just the insect pest. If, for instance, one insect pest were entirely eliminated, there is no knowing what might arise in its place. No one has any idea what effects a given sterilant used on one type of insect will have on other insects, plants, animals, or man for that matter. An action as cruel as ruining and annihilating a family of organisms will surely bring just retribution in its wake.

The aerial spraying of mountain forests with herbicides, pesticides, and chemical fertilizers is considered a success if a given weed or insect pest is selectively killed, or the growth of trees improved, but this is a grave error that can prove most dangerous. Natural conservationists have already recognized such practices as polluting.

Spraying herbicides such as PCP does more than just kill weeds. This also acts as a bactericide and fungicide, killing both black spot on living plants and the many putrefactive fungi and bacteria on fallen leaves. Lack of leaf decomposition seriously affects the habitats of earthworms and ground beetles, on top of which PCP also slaughters microorganisms in the ground.

Treating the soil with chloropicrin will temporarily alleviate bacterial soft rot in Chinese cabbages and the *daikon* radish, but the disease breaks out again two years later, and gets completely out of hand. This germicide halts the soft rot, but at the same time it also kills other bacteria that moderate the severity of the disease, leaving the field open to the soft rot bacteria. This soil germicide also works against fusarium fungi and sclerotium fungi that attack young seedlings, but one cannot overlook the fact that these fungi kill other important pathogens. Is it really possible to restore the balance of nature by spraying an array of bactericides and fungicides like this into a soil populated with such a large variety of microbes?

Instead of trying to bring nature around to his own designs with pesticides, man would be much wiser to step out of the way and let nature carry on its affairs without his interference.

Man is also kidding himself if he thinks that he can clear up the problem of weeds with herbicides. He only makes things harder on himself because this leaves hardy weeds resistant to herbicides or results in the emergence of totally unmanageable new strains of weed. Somebody has come up with the bright idea of killing off herbicide-resistant weeds such as Kentucky bluegrass that are spreading from road embankments by importing an insect pest that attacks weeds. When this insect begins to attack crops, a new pesticide will have to be developed, setting into motion another vicious cycle.

To illustrate just how complex the interrelationships between insects, microorganisms, and plants are, let us take a look at the pine rot epidemic spreading throughout Japan.

The Root Cause of Pine Rot: Contrary to the accepted view, I do not think that the primary cause of the red pine disease that has afflicted so many forested areas of Japan is the pinewood nematode. Recently a group of pesticide researchers at the Institute of Physical and Chemical Research pointed to a new type of *aohen-kin* ("blue mold") as the real villain, but the situation is more complicated than this. I have made a number of observations that throw some light on the true cause.

1. On cutting down a healthy-looking pine in an infected forest, new pathogenic fungi can be isolated from pure cultures of some forty percent of the trunk tissue. The isolated fungi include molds such as *kurohen-kin* ("black mold") and three types of *aohen-kin*, all of them new, undocumented pathogens foreign to the area.

2. Nematode infestation can be observed under a microscope only after a pine is a quarter- or half-withered. Actually, the new pathogenic fungi arrived before the nematodes, and it is on them that the nematodes are feeding, not the tree.

3. The new pathogenic fungi are not strongly parasitic, attacking only weakened or physiologically abnormal trees.

4. Wilting and physiological abnormalities of the red pines are caused by decay and blackening of the roots, the onset of which has been observed to coincide with the death of the *matsutake* mushroom, a symbiont that lives on the roots of red pines.

5. The direct cause of the death of *matsutake* mushrooms was the proliferation of *kurosen-kin* ("bristle mold"), a contributing factor for which was the increasing acidity of the soil.

That red pine disease is not caused by just one organism became clear to me from the results of experiments I conducted on healthy trees in which I inoculated nematodes directly into pines and placed long-horned beetles on the trees under a netting, all without ill effect, and from observations that even when all insect pests are kept off the tree, the roots continue to rot, causing the tree to die. *Matsutake* mushrooms die when small potted pine saplings are placed under conditions of extreme dryness and high temperature, and perish with one hour of exposure at a temperature of 30°C in a hothouse. But they do not die in alkaline soil by the shore with fresh water nearby, or on high ground at low temperature.

On the assumption that red pine disease is triggered first by acidification of the soil and dying of the *matsutake* mushroom, followed by parasitic attack by *kurohen-kin* and other mold fungi, and finally nematode infestation, I tried the following method of control.

1. Application of lime to reduce soil acidification (in the garden, spraying with water containing bleaching powder).

2. Spraying of soil germicides; in gardens, the use of hydrogen peroxide solution and alcohol chloropicrin disinfection is also okay.

3. Inoculation of *matsutake* spores grown in pure culture to promote root development.

These are the bare bones of my method of fighting pine disease, but what most troubles me now is that, although we may be confident of our ability to restore garden trees and cultivate *matsutake* artificially, we are powerless to rehabilitate an ecosystem that has been disturbed.

It is no exaggeration to say that Japan is turning into a barren desert. The loss of the small autumn *matsutake* means more than just the perishing of a mushroom; it is a solemn warning that something is amiss in the world of soil microbes. The first telltale sign of a global change in weather patterns will probably appear in microorganisms. Nor would it be surprising if the first shock wave occurred in the soil where all types of microorganisms are concentrated, or even in mycorrhiza such as *matsutake*, which form a highly developed biological community with very organic interactions.

Essentially, the inevitable happened where it was meant to happen. Red pine is a hardy plant capable of growing even in deserts and on sandy beaches. At the same time, it is an extremely sensitive species that grows under the protection of a very delicate mycorrhiza. Man's ability to control and prevent red pine disease may be a litmus test of his capacity to halt the global loss of vegetation.

3. How Should Nature Be Perceived?

Seeing Nature as Wholistic

The central truth of natural farming is that nothing need be done to grow crops. I have learned this because non-discriminating knowledge has enabled me to confirm that nature is complete and crops more than capable of growing by themselves. This is not the theoretical hypothesis of a scholar in his study or the wishful thinking of an idler with an aversion to work; it is based on a total, intuitive understanding of the truth about self and nature wrested from the depths of doubt and skepticism in a deeply earnest struggle over the meaning of life. This is the source of my insistence that nature not be analyzed.

Examining the Parts Never Gives a Complete Picture: This is extremely important, but since it is somewhat abstract, I will illustrate with an example.

A scientist who wishes to know Mt. Fuji will climb the mountain and examine the rocks and wildlife. After having conducted geological, biological, and meteorological research, he will conclude that he now has a full picture of Mt. Fuji. But if one were to ask whether it is the scientist who has spent his life studying the details of Mt. Fuji who knows it best, the answer would have to be no. When one seeks full understanding and comprehensive judgment, analytic research is instead a hindrance. If a lifetime of study leads to the conclusion that Mt. Fuji consists mostly of rocks and trees, then it would have been better not to have climbed it in the first place.

One can know Mt. Fuji by looking at it from afar. One must see it and yet not examine it, and in not examining at it, know it.

Yet the scientist will think: "Well, gazing at Mt. Fuji from a distance is useful for knowing it abstractly and conceptually, but is no help in learning something about the actual features of the mountain. Even if we concede that analytic research is of no use in knowing and understanding the truth about Mt. Fuji, learning something about the trees and rocks on the mountain is not totally meaningless. And moreover, isn't the only way to learn something to go and examine it directly?"

To be sure, I can say that to analyze nature and append to these observations one's conclusions is a meaningless exercise, but unless those who listen understand why this is worthless and unrelated to the truth, they will not be convinced.

What more can I say if, when I mention that the artist Hokusai who captured faraway images of Mt. Fuji in his paintings understood it better than those who climbed it and found it an ugly mountain, I am told that this is just a subjective difference, a mere difference in viewpoint or opinion.

The most common view is that one can best know the true nature of Mt. Fuji by listening to the ecologist speak of his research on its fauna and flora and by looking at the abstracted form of Mt. Fuji in Hokusai's paintings. But this is just like the hunter who chases two rabbits and catches none. Such a person neither climbs the mountain nor paints. Those who say Mt. Fuji is the same whether we look at it lying down or standing up, those who make use of discriminating knowledge, cannot grasp the truth of this mountain.

Without the whole, the parts are lost, and without the parts, there is no whole. Both lie within the same plane. The moment he distinguishes between the trees and rocks that form a part of the mountain and the mountain as a whole, man falls into a confusion from which he cannot easily escape. A problem exists from the moment man draws a distinction between partial, focused research and total, all-encompassing conclusions. To know the real Mt. Fuji, one must look at the self in relation to Mt. Fuji rather than at Mt. Fuji.

One must look at oneself and Mt. Fuji prior to the self-other dichotomy. When one's eyes are opened by forgetting the self and becoming one with Mt. Fuji, then one will know the true form of the mountain.

Become One with Nature: Farming is an activity conducted by the hand of nature. We must look carefully at a rice plant and listen to what it tells us. Knowing what it says, we are able to observe the feelings of the rice as we grow it. However, "to look at" or "scrutinize" rice does not mean to view rice as the object, to observe or think about rice. One should essentially "put oneself in the place of the rice." In so doing, the self looking upon the rice plant vanishes. This is what it means to "see and not examine and in not examining to know." Those who have not the slightest idea of what I mean by this need only devote themselves to their rice plants. It is enough to be able to work with detachment, free of worldly concerns. Leaving aside one's ego is the quickest path to unity with nature.

Although what I am saying here may seem as intangible and difficult to understand as the words of a Zen priest, I am not borrowing philosophical and Buddhist terms to spout empty theories and principles. I am speaking from raw personal experience of things grounded solidly in reality.

Nature should not be taken apart. The moment it is broken down, parts cease being parts and the whole is no longer a whole. When collected together, all the parts do not make a whole. "All" refers to the world of mathematical form and "whole" represents the world of living truth. Farming by the hand of nature is a world alive, not a world of form.

The instant he begins to ponder over the factors of crop cultivation and growth and concerns himself with the means of production, man loses sight of the crop as a whole entity. To produce a crop, he must comprehend the true meaning of a plant growing on the earth's surface, and the goal of production must derive from a clear vision of unity with the crop.

Natural farming is one way to remedy the presumptions and conceits of scientific thought, which claims to know nature and says man produces crops. Natural farming checks whether nature is perfect or imperfect, whether it is a world of contradiction. The task then is to establish and prove whether pure natural farming free of all vestige of the human intellect is indeed powerless and inferior, and whether farming based on the inputs of technology and scientific knowledge is truly superior.

For several decades now, I have devoted myself to examining whether natural farming can really compete with scientific farming. I have tried to gauge the strength of nature in rice and barley cultivation, and in the growing of fruit trees. Casting off human knowledge and action, relying only on the raw power of nature, I have investigated whether "do-nothing" natural farming can achieve results equal to or better than scientific farming. I have also compared both approaches using man's direct yardsticks of growth and yield. The more one studies and compares the two, whether from the limited perspective of growth and yields, or from a broader and higher perspective, the clearer and more undeniable becomes the supremacy of nature.

However, my research on natural farming has done more than just point out the faults of scientific farming, it has given me a glimpse of the disasters that the frightening defects of modern practices are visiting on mankind.

Imperfect Human Knowledge Falls Short of Natural Perfection: Understanding the degree to which human knowledge is imperfect and inadequate helps one to appreciate just how perfect nature is. Scientists of all ages have sensed with increasing clarity the frailty and insignificance of human knowledge as man's learning grew from his investigations of the natural world around him. No matter how unlimited his knowledge may appear, there are hurdles over which man cannot pass: the endless topics that await research, the infinitude of microscopic and submicroscopic universes that even the rapid specialization of science cannot keep pace with, the boundless and eternal reaches of outer space. We have no choice then but to frankly acknowledge the frailty and imperfection of human knowledge. Clearly, man can never escape from his imperfection.

If human knowledge is unenlightened and imperfect, then the nature perceived and built up by this knowledge must in turn always be imperfect. The nature perceived by man, the nature to which he has appended human knowledge and action, the nature which serves as the world of phenomenon on which science acts, this nature being forever imperfect, then that which is opposed to nature—that which is unnatural, is even more imperfect.

And paradoxically, the very incompleteness of the nature conceived and born of human knowledge and action—a nature that is but a pale shadow of true nature—is proof that the nature from which science derived its image of nature is whole and complete.

The only direct means for confirming the perfection of nature is for each individual to come into immediate contact with nature in its true form and see for himself. People must experience this personally and choose to believe or not believe.

As for me, I have found nature to be perfect and am trying here only to present the evidence. Natural farming begins with the assumption that nature is perfect.

Natural farming starts out with the conviction that barley seeds which fall to the earth will send up sprouts without fail. If a barley sprout should emerge then later wilt in mid-growth, something unnatural has occurred and one reflects on the cause, which originates with human knowledge and action. One never blames nature, but begins by blaming oneself. One searches unrelentingly for a way to grow barley in the heart of nature.

There is no good or evil in nature. Natural farming admits to the existence neither of insect pests nor of beneficial insects. If a pest outbreak occurs, damaging the barley, one reflects that this was probably triggered by some human mistake. Invariably, the cause lies in some action by man; perhaps the barley was seeded too densely or a beneficial fungus that attacks pests was killed, upsetting nature's balance. Thus, in natural farming, one always solves the problem by reflecting on the mistake and returning as close to nature as possible.

Those practicing scientific farming habitually blame insect infestation on the weather or some other aspect of nature, then apply pesticides to exterminate the marauding pest and spray fungicides to cure diseases.

The road diverges here, turning back to nature for those who believe nature to be perfect, but leading on to the subjugation of nature for those who doubt its perfection.

Do Not Look at Things Relatively

In natural farming, one always avoids seeing things in relative terms; should one catch sight of relative phenomena, one immediately tries to trace these back to a single source, to reunite the two broken halves.

To farm naturally, one must question and reject scientific thinking, all of which is founded on a relative view of things: notions of good and poor crop growth, fast and slow, life and death, health and disease, large and small yields, large and small gains, profits and losses.

Let me now describe what constitutes a viewpoint that does not fall prey to relativistic perceptions so that I may help correct the errors committed by a relative view of things.

From a scientific perspective, things are large or small, dead or alive, increasing or decreasing. But this view is predicated on notions of time and space, and is really nothing more than a convenient assumption. In the natural world, which transcends time and space, there is, properly speaking, no large or small, no life or death, no rise or fall. Nor was there ever the conflict and contradiction of opposing pairs: right and left, fast and slow, strong and weak.

If we go beyond the confines of time and space, we see that the autumn death of a rice plant can be understood as life passing into the seed and continuing on into eternity. Only man frets over life and death, gain and loss. A method of farming founded on the view of birth as the beginning and death as the end cannot help but be short-sighted.

In the narrow scientific view, growth appears to be either good or poor, and yields either large or small, but the amount of sunlight reaching the earth stays constant and the levels of oxygen and carbon dioxide remain balanced in the atmosphere. This being so, why do we nevertheless see differences in growth and yields? The fault is usually man's. Man destroys the immutability and stability of nature either by himself invoking the notions of large and small, many and few, or by altering form and substance. These things become self-evident when viewed from a deeper and broader perspective or from a perspective in accordance with nature.

Man generally finds value only in the harvest of grains and fruit. But nature sees both cereal grains and weeds, and all the animals and microorganisms that inhabit the natural world, as the fruit of the earth. Notions of quantity and size usually exist within a limited frame of reference. From a broader or slightly more relaxed perspective, these cease to be problems at all.

When looking at nature from the standpoint of natural farming, one does not worry over minor circumstances; there is no need for concern over form, substance, size, hardness, and other peripheral matters. Such concerns only cause us to lose sight of the real essence of nature and shut off the road back to nature.

Take a Perspective That Transcends Time and Space

I have said that to travel the road leading to a natural way of farming, one must reject the use of discriminating knowledge and not take a relativistic view of the world. These may be thought of as means for attaining a perspective transcending time and space. A world without discrimination, an absolute world that passes beyond the reaches of the relative world is a world that transcends space and time.

When captive to the notions of space and time, we are capable only of seeing things circumstantially. Scientific farming is a method of farming that originates within the confines of time and space, but Mahayana natural farming comes into being only in a world that transcends time and space.

Thus, in striving to realize a natural way of farming, one must focus one's efforts on overcoming time and space constraints in everything one does. Transcending time and space is both the starting point and the destination of natural farming. Scientific farming, with its concerns for harvesting so much from a given field over such-and-such a period of time, is confined within the limits of time and space, but in natural farming one must go beyond space and time by making decisions and achieving results from a general, not local, perspective and fortified by a position of freedom and a long-term viewpoint.

To give an example, when an insect alights on a rice plant, science immediately zeros in on the relationship between the rice plant and the insect; if the insect feeds on juices from the leaf of the plant and the plant dies, then the insect is viewed as a pest. Research is carried out on the insect, which is identified taxonomically, and its morphology and ecology studied carefully. This knowledge is eventually used to determine how to kill the pest.

The first thing that the natural farmer does when he sees this crop and the insect is to see, yet not see, the rice; to see and yet not see the insect. He is not misled by circumstantial matters; he does not pursue the scientific method of inquiry by observing the rice and insect or investigating what the insect is; he does not ask why, when, and from where it came, or try to find out what it is doing in his field. What then does he do? He reaches beyond time and space by taking the stance that there are no crops or pests in nature to begin with. The concepts of "raising plants" and "harmful insects" are just words coined by man based on subjective criteria grounded in the self; viewed in terms of the natural order, they are meaningless. This insect is thus a pest and yet not a pest. Which is to say that its presence in no way interferes with growth of the rice plant for there is a way of farming in which both the rice plant and insect can coexist in harmony.

Natural farming seeks to develop methods of rice cultivation in which the existence of "pests" poses no problem. It begins by first stating the conclusion and clearing up local and temporal problems in a way that fits the conclusion. Even leafhoppers, pests from the scientific viewpoint, do not always harm rice. The time and circumstances also play a part.

When I say that it is necessary to examine things from a broad, long-range perspective, I do not mean that one must conduct difficult and highly specialized research.

The scientist studies rice damage by a particular insect, but it would suffice to observe cases where the insect does no damage to the rice. Such cases invariably exist. Instances of damage are quite naturally accompanied also by instances of no damage. There may be immense damage in one field and none in another. Invariably too, there are cases in which the insects will not even approach the rice. Natural farming examines cases in which little or no damage occurs and the reasons why, based on which it creates circumstances where nothing is done, yet insect damage is nonexistent.

One type of leafhopper that attacks rice plants early in the growing season is the green rice leafhopper, which lives among the weeds in the levees between rice fields from winter to early spring. To rid the fields of these leafhoppers, burning the levee weeds is preferable to direct application of a leafhopper poison. But an even better way is to change the variety of weeds growing on the levees.

The white-backed leafhopper that emerges in the summer and the brown leafhopper, large outbreaks of which sometimes occur in the fall, tend to appear during long spells of hot, humid weather, but break out in especially large numbers in the summer or fall in flooded fields of stagnant water. When the field is drained and the surface of the field exposed to breezes so that it dries, spiders and frogs emerge in number, helping reduce damage to a minimum.

The farmer need not worry about damage by leafhoppers if he cultivates healthy fields of rice. Nature is always showing man, at some time and place, situations in which pests are not pests and do not cause real damage. Instead of holing up in laboratories, people can learn directly in the open classrooms of nature.

Natural farming takes its departure from a perspective transcending time and space, and returns to a point beyond time and space. Man must learn from nature the bridge that links these two points. The real meaning of taking a transcendent perspective, in plain, down-to-earth terms, is to help provide both insect pests and beneficial insects with a pleasant environment in which to live.

Do Not Be Led Astray by Circumstance

To look at things from a perspective that transcends time and place is to prevent oneself from becoming captive to circumstance. Even science constantly tries to avoid becoming too wrapped up in details and losing sight of the larger picture. However, this "larger picture" is not the true picture. There is another view that is broader and more all-encompassing.

In nature, a whole encloses the parts, and a yet larger whole encloses the whole enclosing the parts. By enlarging our field of view, what is thought of as a whole becomes, in fact, nothing more than one part of a larger whole. Yet another whole encloses this whole in a concentric series that continues on to infinity. Therefore, while it can be said that to act one must intuitively grasp the true "whole" and include therein all small particulars, this cannot actually be done.

Let us take an example from the world of medicine. The physician studies the stomach and intestines, examines the ingredients of various foods, and investigates how these are absorbed as nutrients by the human body. The common perception is that, as research becomes increasingly focused and parallel advances are made in broad, interdisciplinary studies, nutritional science becomes an authoritative field in its own right with application to all cases.

But for all we know, nutritional science, which was introduced to Japan from Western Europe, may have first been modeled on German beer drinkers or French wine lovers. Nutritional principles that work for them do not necessarily apply to the people of Africa, for example. The same *daikon* will be absorbed very differently and will have an entirely different nutritive value for the irritable city dweller afflicted by smog and noise pollution who eats his *daikon* without secreting digestive juices, as compared with the tropical African who munches on his after a meal of wild game.

Progress in medicine has brought us a whole host of dietary therapies, such as low-calorie diets for people who want to lose weight, light diets for people with stomach problems, low-salt diets for people with bad kidneys, and sugarless diets for people with pancreatic ailments. But what happens when a person has problems with two or three organs? If this food is out and that one forbidden, then the poor fellow, unable to eat anything, could end up as thin as a dried sardine.

It is a mistake to believe that as advances are made in a broad range of highly specialized fields, the scope of applications grows. We should not forget that the more highly specialized the research, the further it strays from a broad perspective.

In an age before the development of nutritional science, before we gave any thought about what was good or bad for us, all we knew was that to stay healthy, one should eat in moderation. Which has broader application, which is more effective: modern nutritional science with its specialized research or traditional admonitions for moderation at mealtime? Modern nutritional science may appear to have broader application because it considers all cases, yet it forbids first one thing then another, so people keep running into walls and struggling with a lot of new problems. Cruder but complete, the simple knowledge that one should eat with moderation applies to all people and so it works better. This is so because knowledge that is less discriminating has wider application.

Be Free of Cravings and Desires

The aim of scientific farming is to chase after the objects of man's desire, but natural farming does not seek to satisfy or promote human cravings. Its mission is to provide the bread of human life. This is all it seeks, no more. It knows how much is enough. There is no need to become caught up in man's cravings and attempt to expand and fortify production.

What has the campaign in Japan to produce good-tasting rice over the last several years achieved? How much happier does it make us when a farmer throws himself into improving varieties and raising production in response to the vagaries of the consumer for "tasty" rice and barley. Only the farmer suffers, because nature strongly resists all his efforts to upgrade crops for minor gains in taste and sweetness. Do urbanites know the torments that farmers go through—declines in production, reduced crop resistance to diseases and pests, to give but a couple examples—when consumers demand the slightest improvement in flavor?

Nature sounds warnings and resists man's unnatural demands. Only, it says nothing. Man must make reparations for his own sins. But he cannot forget the sweetness he has tasted. Once the cravings of the palate assert themselves, there is no retreating. No matter how great the labors that farmers must shoulder as a result, these are of no concern to the consumer. Scientific farming exalts and follows the example of the farmer working diligently to service the endlessly growing demands of city dwellers, who expect, as a matter of course, fresh fruit and beautiful flowers in all seasons.

The fruits of autumn picked in the fields and mountains were beautiful and sweet. The beauty of flowers in the meadow was a thing to behold. Natural farming tries to enter the bosom of nature, not break it down from without; it has no interest in conquering nature, but seeks instead to obey it. It serves not man's ambitions, but nature, reaping its fruit and wine. To the selfless, nature is always beautiful and sweet, always constant. Because all is essentially one.

No Plan Is the Best Plan

If nature is perfect, then man should have no need to do anything. But nature, to man, appears imperfect and riddled with contradiction. Left to themselves, crops become diseased, they are infested by insects, they lodge and wither.

But upon taking a good look at these examples of imperfection, we realize that they occur when nature has been thwarted, when man has fiddled with nature. If nature is left in an unnatural state, this inevitably invites failure, leading not only to imperfection, but even catastrophe.

When nature appears imperfect this is the result of something man has done to nature that has never been rectified. When left to its proper cycles and workings, nature does not fail. Nature may act, or may compensate or offset one thing for another, but it always does so while maintaining order and moderation.

The pine tree that grows on a mountain rises up straight and true, sending out branches in all directions in a regular annular pattern. In keeping with the rule of phyllotaxy, the branches remain equally spaced as they grow, so no matter how many years pass, branches never crisscross or overlap and die. The tree grows in just the right way to allow all the branches and leaves to receive equal amounts of sunlight.

But once a pine has been transplanted into a garden and pruned with clippers, the arrangement of branches undergoes a dramatic change, taking on the contorted "elegance" of a garden tree. This is because, once it has been pruned, a pine no longer sends out normal shoots and branches. Instead, branches grow irregularly, crisscrossing every which way, bending, twisting, and overlapping with each other. By merely nipping the buds at the tips of citrus shoots, conical citrus trees that had until then grown straight fork into a three-leader arrangement or assume a wineglass shape. The same is true of all trees.

Once man comes into the act, a tree loses its natural form. In a tree of unnatural habit, the branches are in disarray, growing either too close together or too far apart. Diseases arise and insects burrow and nest wherever there is poor ventilation or inadequate exposure to sunlight. And where two branches cross, a struggle for survival ensues; one will thrive, the other die. All it takes to destroy the conditions of nature and transform a tree that lived in peace and harmony into a battleground where the strong consume the weak is to nip a few young buds.

Although disruption of the order and balance of nature may have begun as the unintentional consequence of impulsive human deeds, this has grown and escalated to the point where there is no turning back. Once tampered with, the garden pine can never revert back again to being a natural tree. All it takes to disturb the natural habit of a fruit tree is to nip a single bud at the end of a young shoot.

When nature has been tainted and left unnatural, what remains? Here begins the never-ending toil of man. Two crisscrossing branches compete with each other. To prevent this, man must meticulously prune the garden pine each year.

Snipping off the tip of a branch causes several irregular branches to grow in its place. The tips of these new branches must then be cut the next year. The following year, the even larger number of new branches create even greater confusion, increasing the amount of pruning that has to be done.

The same holds true for the pruning of fruit trees. A fruit tree pruned once must be tended for its entire life. The tree is no longer able to space its branches properly and grow in the direction it chooses. It leaves the decision up to the farmer and just sends out branches wherever and however it pleases without the least regard for order or regularity. Now it is man's turn to think and cut the branches not needed. Nor can he overlook those places where the branches cross or grow too densely together. If he does, the tree will grow confused; branches at the center will rot and wither, and the tree will become susceptible to disease and insects and eventually die.

Man is compelled to act because he earlier created the very conditions that now require his action. Because he has made nature unnatural, he must compensate for and correct the defects arising from this unnatural state.

Similarly, man's deeds have made farming technology essential. Plowing, transplanting, tillage, weeding, and disease and pest control; all these practices are necessary today because man has tampered with and altered nature.

The reason a farmer has to plow his rice field is that he plowed it the year before, then flooded and harrowed it, breaking the clods of earth into smaller and smaller particles, driving the air out and compacting the soil. Because he kneads the earth like bread dough, the field *has* to be plowed each year. Naturally, under such conditions, plowing the field raises productivity.

Man also makes crop disease and pest control indispensable by growing unhealthy crops. Agricultural technology creates the causes that produce disease and pest damage, then becomes adept at treating these. Growing healthy crops should take precedence.

Scientific farming attempts to correct and improve on what it perceives as the shortcomings of nature through human effort. In contrast, when a problem arises, natural farming relentlessly pursues the causes and strives to correct and restrain human action.

The best plan, then, is true non-action and no plan at all.

4.   Natural Farming for a New Age

At the Vanguard of Modern Farming

To some, natural farming may appear as the return to a passive, primitive form of farming over the road of idleness and inaction. Yet because it occupies an immutable and unshakable position that transcends time and space, natural farming is always both the oldest and the newest form of farming. Today, it presses on at the very leading edge of modern agriculture.

Although the truth remains fixed and immobile, the heart of man is ever fickle and changing; his thinking shifts with the passage of time, with circumstances, and so he is forced to alter his means. He, and science with him, orbits forever about the periphery without reaching in to the truth at the center.

Scientific farming blindly traces spiraling cycles in the tracks of science. Today's new technology will become the dated technology of tomorrow, and tomorrow's reforms will become the stale news of a later day. What is on the right today will appear on the left tomorrow, and on the right the day after. While this wheel spins round and round, it expands and diffuses outward.

Even so, things were better when man circled about the periphery while gazing from afar upon the truth at the center. Man today tries to leap outside of nature and truth altogether. Balanced against this centrifugal force are the centripetal forces, represented by efforts to return to nature and to see the truth, that have managed only barely to maintain a balance. But the moment this thread connected to the core breaks, man will fly away from truth like a whirling stone. The danger has now arrived at the doorstep of science. Scientific farming has no future.

Natural Livestock Farming

The Abuses of Modern Livestock Farming: The storms of agricultural reform are beginning to ravage the good name of agricultural modernization. Let us look at a trend that has emerged in all farming technologies.

One new livestock technology that has been spreading like wildfire throughout Japan is the mass raising of of chickens, pigs, cattle, and other livestock and fowl in large facilities. The animals are fed preserved foods compounded from a very small amount of natural feed and liberal amounts of additives such as drugs, vitamins, and nutrients, all ostensibly for protecting health. This eliminates the necessity of rushing about to attend to every need of the livestock. The animal is efficiently raised by placing it in a narrow enclosure or cage just big enough to accommodate it but hardly allowing it to move about. The goal is to produce as much as possible on a narrow piece of land.

There appear to be no problems with this method. In addition to being efficient, the work is less physically demanding and production is better than ever. But high-volume livestock farming encounters the problems of market supply and distribution of the product familiar in factory production. Beset by wildly fluctuating prices, the livestock farmer becomes totally caught up with concerns over his margins and profits.

The quality of these products is in every way inferior to beef and eggs from cattle and fowl allowed to gambol and play freely outdoors, and to multiply and grow without restraint. What's more, because these animals have been raised on roughage packed with antibiotics, preservatives, flavor enhancers, hormones, and residual pesticides, there is also the concern that toxins harmful to the human body have accumulated in the beef and eggs. We have arrived in an age where beef is no longer beef and eggs are no longer truly eggs. What we have instead is merely the conversion of a complete feed preparation into meat or eggs. Livestock farming is no longer a form of agriculture practiced in nature. Unfertilized battery chickens are just machines for hatching factory-made eggs, and pigs and cows are merely factory-produced meat and milk-fabricating machines. These products could not possibly be wholesome. The point is that, regardless of whether the product is good or bad, one person can raise tens and hundreds of thousands of head efficiently with mass production techniques. But it is capital, not men, that today raises these animals. This is no longer the farmer's domain, but that of commercial houses which raise livestock in large factory-like operations.

Natural Grazing Is the Ideal: Is natural livestock farming old and outdated in contrast? Under the precepts of natural farming, livestock farming takes the form of open grazing. Cattle, pigs, and chickens fattened while free to roam at will on the open land under the sun's rays are a precious, irreplaceable source of food for man. The problem lies elsewhere—in the prejudiced view that sees natural farming as inefficient. Is grazing, which allows one person to raise hundreds of head without doing anything, really inefficient? Is it not, rather, the most efficient form of production there is?

This is not to say that raising cattle, pigs, and chickens freely in open meadows and forests is without its problems. There are poisonous plants, diseases, and ticks. Some would even call free grazing unhygienic. But most such problems are the consequence of human action and can be resolved. The basic premise that animals are perfectly capable of being born and living in nature is unassailable, and so, although solutions may require some very determined observation, there is always a way. The key is to raise the right animal in the right environment while letting nature be.

Even fields covered with a thick growth of wild roses and creepers that seem worthless for grazing can be used to raise goats and sheep, which love to feed on these intractable shrubs and vines and could clean up the undergrowth in the densest jungle.

There is no need to worry that cows or other animals cannot be raised in uncultivated pastures. They can be raised in mixed woods or even in mountain forests planted with Japanese cypress or pine. Grasses and underbrush have to be cut the first seven or eight years after planting trees on a mountain, but the trouble of cutting the brush can be eliminated very nicely by raising cows. The grazing cattle may slightly damage a few young saplings along a fixed path through the cypresses, but the planted saplings will remain almost entirely un-affected. This may seem hard to believe, but it is only natural when we recall that animals in nature do not indiscriminately ravage anything unrelated to what they eat. Obviously, a natural forest would be even more ideal than a reforested area.

In allowing animals to graze in the fields and mountains, some people may worry about the presence of poisonous plants, but animals have an innate ability to tell these apart from other plants. If no longer able to do so, there is most certainly a reason why. Bracken, for example, may be a poisonous herb under certain conditions, but it grows in clusters. If a cow eats too much and suffers, something is probably wrong with the cow.

Livestock bred by artificial insemination and raised on artificial milk formulas are more likely to have poor viability. Animals improved indiscriminately often show unanticipated defects. Genetic improvement programs are usually opposed to nature, and often result in the creation of unnaturally deformed creatures that man deludes himself into thinking are superior.

It would be unreasonable, of course, to take modern, genetically upgraded livestock, release them suddenly in a forest, and expect to see an immediate improvement in results. But if the possibilities are studied with patience, a path should open up. At the very least, after habituating the animals to open grazing in the forests over the course of two or three generations, natural selection will take over and those animals adapted to nature will survive.

Ticks and mites do present a problem, but the conditions under which parasites such as these arise vary considerably. There may be a great number at the southern edge of the woods, but very few along the northern edge. Infestation is generally limited in cool, breezy areas, and is closely related to humidity and temperature. The problem can be prevented by providing the right environment. It should suffice to raise hardier cattle and give some consideration to the protection and raising of beneficial insects that help control the tick population.

It will also be necessary to stop thinking in terms of raising just cattle. What happens, for example, when we let pigs, chickens, and rabbits graze together with the cows in an orchard? The pigs like to root up the ground looking for insects and earthworms they are fond of in valleys and damp areas; they are like small tractors that dig up the soil. Just sow some clover and grain in the turned soil, and with the cow and pig droppings, you should get a fine growth of pasturage. Once this pasture grass begins to flourish, then you should be able to raise chickens, goats, and rabbits in the same way.

Today's livestock, raised in large numbers and reduced to just so much standardized machinery, no longer receives the strength or blessings of nature. As the products of human endeavor through the power of science alone, they differ fundamentally from nature, which creates something from nothing, because they are merely processed goods, the transformation of one thing into another.

Livestock production under factory-like conditions is generally thought to be efficient, but this is a near-sighted assessment based on a limited spatial and temporal frame of reference. The pitiful sight of fowl, pigs, and cattle confined to cages and unable even to move bears witness to the loss of nature of these animals and points also to man's alienation to and loss of nature. Both the farm worker directly engaged in the raising of livestock and the city dweller who consumes these food products lose their health and their humanity as they turn away from nature.

Livestock Farming in the Search for Truth: Scientific farming is satisfied to think of conditional truth as *the* truth, but natural farming makes every effort to discard all premises and conditions and seek out a truth without conditions.

For instance, when studying a particular animal feed, scientific farming will give various formulations to cows chained to a barn (representing a certain set of environmental conditions), and judge the mixture producing the best results to be superior to the others (inductive experimentation). From this, it draws various conclusions about cattle feed, which it believes to be the truth.

Natural farming does not follow this type of reasoning and experimental approach. Because its goal is unconditional truth, it begins by examining the cow from a standpoint that disregards environmental conditions, by asking how the cow lives in open nature. But it does not immediately analyze what the cow eats when and where. Rather, it takes a broader perspective and looks at how a cow is born and grows. By paying too much attention to what the cow feeds on, we lose a broader understanding of how it lives and what it needs to live. More is needed to sustain life than just food. Nor are problems of sustenance resolved by food alone. Many other factors relate to life: weather, climate, living environment, exercise, sleep, and more. Even on the subject of food, what a cow does not eat, dislikes, or has low nutritive value is generally thought worthless, but may actually be indispensable in certain cases. We must therefore find a way, within the broad associations between man, livestock, and nature, of rearing livestock that leaves the animals free and unrestrained.

The very notion of "raising" livestock should not even exist in natural farming. Nature is the one that raises and grows. Man follows nature; all he needs to know is how and why cattle live. When he designs and builds a barn or a chicken coop, a farmer should not rely on his human reasoning and feelings. Even when the scientist conducts separate studies on such factors as temperature and ventilation and runs experiments in which he raises calves or chicks under given conditions, it is only natural that his results will show that these should be raised under cool conditions in summer and warm conditions in winter. The conclusion (scientific truth) that an optimum temperature is needed to raise the calves or chicks is a natural consequence of the method used to raise these, and certainly is not an immutable truth.

Although high and low temperatures exist in nature, the notions of hot and cold do not. Although cattle, horses, pigs, sheep, chickens, and ducks all know the difference between hot and cold, they never complain to nature that it is hot or cold. With our temperate climate in Japan, there never was a need to worry about whether the summer heat or winter cold was good or bad for raising animals. And there certainly was no call for the frantic concern over whether they would live or die.

Heat and cold exist, and yet do not exist, in nature. One will never be wrong in starting with the assumption that the temperature and humidity are everywhere and at all times just right. The size, height, frame, construction, windows, floor, and other features of animal enclosures have been improved on the basis of diverse theories, but we have to return to the starting point and try making a fundamental turnabout. Without hot and cold, the barn is no longer necessary. All that is needed, for the convenience of man, is the smallest of sheds: perhaps a shed in which to milk the cows and a tiny chicken shed in which hens can lay their eggs. As for the animals, they will scratch and forage freely for food night and day under the open sky, find themselves a place to roost, and grow up strong and healthy. Lately, disease has become a frequent problem in animal husbandry and because it is often a major factor in determining whether a livestock operation will succeed or fail, farmers are racking their brains to find a solution. This problem too will never really be solved unless farmers make their starting point the raising of healthy animals that do not contract diseases.

Some eighty percent of Japan consists of mountains and valleys. One could probably fence off the entrance to one of those depopulated mountain villages that have lost their inhabitants to the cities and thus create a large, open grazing range for animals. I would like to see someone try an experiment on this scale. All sorts of domestic animals could be placed inside the enclosure and left to themselves for a number of years, after which we could go in and see what had happened.

To summarize, then, scientific experiments always take a single topic and subject it to a number of variable conditions while making some prior assumption about the results. Natural farming, however, pushes aside all conditions, and knocking away the precepts from which science operates, strives to find the laws and principles in force at the true source.

Unchanging truths can be found only through experiments free of conditions, assumptions, and notions of time and space.

Natural Farming—In Pursuit of Nature

There is a fundamental difference between nature and the doctrine of laissez-faire or non-intervention. Laissez-faire is the abandoning of nature by man after he has altered it, such as leaving a pine tree untended after it has been transplanted in a garden and pruned, or suddenly letting a calf out to pasture in a mountain meadow after raising it on formula milk.

Crops and domestic animals are no longer things of nature and so it is already close to impossible to attain true Mahayana natural farming, but at least we can try reaching for Hinayana natural farming, which approaches closest to nature. The ultimate goal of this way of natural farming is to know the true spirit and form of nature. To do this, we can start by closely examining and learning from a laissez-faire situation before us. By observing nature that has been abandoned by man, we can make out the true form of nature that lies behind it. Our goal then is to carefully examine abandoned nature and learn of the true nature revealed when the effects of man's earlier actions are removed.

But this will not suffice to know nature in its true form. Even nature stripped of all human action and influence is still only nature as seen through man's relativity, a nature clothed in the subjective notions of man. To follow the path of natural farming, one must tear the robes of human action from nature and remove the innermost garments of subjectivity.

One must beware also of arbitrarily settling upon causal relationships on the basis of subjective human notions, or of drawing suppositions on the questions of accident and necessity or the association between continuity and discontinuity. One must first follow closely on nature's heels, rejecting all assumptions, knowledge, and action; not thinking, not seeing, not doing. That nature is God.

The Only Future for Man

Will humanity go on advancing without end? The people of this world seem to think that, although reality is rife with contradiction, development will continue forever in a process of sublation while wandering between right and left, and thesis-antithesis-synthesis.

Yet the universe and all it contains does not advance along a linear or planar path. It expands and grows volumetrically outwards and must, at the furthest limit, rupture, split, collapse, and disappear. But at a point beyond this limit, what should have vanished reverses its course and reappears, now moving centripetally inward, contracting and condensing. What has form vaporizes at the limits of development to a void, and the void condenses into a form and reappears, in a never-ending cycle of contraction and expansion.

I liken this pattern of development to the Wheel of Dharma or a cyclone because it is identical to a cyclone or tornado, which compresses the atmosphere into a vortex, expanding and growing as it rages furiously, then eventually disintegrates and vanishes.

Human progress also moves toward collapse. The question is how, and in what manner shall it come to ruin? I have sketched below how I believe this will inevitably occur and what man must do.

The first stage of this collapse will be the breakdown of human knowledge. Human knowledge is merely discriminating knowledge. Having no way of knowing that this knowledge is really unknowable, man founders ever deeper into confusion through the collection and advancement of unknowable and mistaken knowledge. Unable to extricate himself from schizophrenic development, he ultimately brings upon himself spiritual derangement and collapse.

The second stage will be the destruction of life and matter. The earth, an organic synthesis of these two elements, is being broken down and divided up by man. This is gradually depriving the natural world on the earth's surface of its equilibrium. Destruction of the natural order and the natural ecosystem will rob matter and life of their proper functions. Nor will man be spared. Either he will lose his adaptability to the natural environment and meet with self-destruction or he will succumb to instant ruin under a slight pressure from without, like an inflated rubber balloon ruptured by a small needle.

The third stage will be failure, when man loses sight of what he must do. The industrial activity that expands relentlessly with developments in the natural sciences is basically a campaign to promote energy consumption. Its target has not been so much to boost energy production as to senselessly waste energy. As long as man continues to take the stance that he is "developing" nature, the materials and resources of the earth will go on drying up. Burdened by growing self-contradictions, industrial activity will grind to a halt, or undergo unyielding transformations that shall usher in drastic changes in political, economic, and social institutions.

Self-contradiction is most evident in the decline in energy efficiency. In his fascination with ever greater sources of energy, man has moved from the heat of the fireplace to electrical generation with a water wheel to thermal power generation to nuclear power, but he closes his eyes to the fact that the efficiency of these sources (ratio of total energy input to total energy output) has worsened exponentially in the same order. Because he refuses to acknowledge this, internal contradiction continues to accumulate and will soon reach explosive levels.

Some scientists believe that, if nuclear energy dries up, then we should turn to solar energy or wind power, which are non-polluting and do not engender contradictions. But these will only continue the decline in energy efficiency and, if anything, will accelerate the speed at which man heads toward destruction.

Until man notices that scientific truth is not the same as absolute truth and turns his system of values on its head, he will continue to rush blindly onwards toward self-destruction.

There will then he nothing for him to do except sustain an attitude that enables him to survive without doing anything. Man's only work then will consist of the barest of farming essential for sustaining life. But since agriculture does not exist as an independent entity of and for itself, the farming he will practice will not be an extension of modern agriculture.

Farming with small machinery was more energy efficient than modern large-scale agriculture using large implements, and farming with animal power was even more efficient. Properly speaking, no form of agriculture has better energy efficiency than natural farming. Once this becomes clear, people will realize for themselves what they must do.

Only natural farming lies in the future. And natural farming is the only future for man.

Rice

1. Rice seedlings covered with a mulch of barley straw (June).

2. Rice growing in a ground cover of bur clover.

3. Early June.

4. Rice in a ground cover of clover (mid-June).

5. Rice amid the clover (July).

6. August.

7. September.

8. Mature rice (October).

1.

2. December.

3. Barley at the tillering stage, in a mulch of rice straw (February).

4. Barley growing among the clover.

5. Barley sown among the weeds (April).

6. April.

7. May.

8. Preparing clay pellets for sowing rice seed before the barley harvest.

Vegetables

1. Flowering vegetables growing by the roadside.

2. Flowering *daikon*.

3. These *daikon* and turnips were grown without fertilizers or pesticides.

4. A *daikon* doing a loop-the-loop among the clover.

5. Mixture of potherb mustard and other vegetables.

6. *Daikon*, turnips, and other vegetables growing together.

7. Six hundred fruits from a single hill of chayote.

8. *Daikon* in full flower.

Fruit Trees

1. The view from my hilltop citrus orchard.

2. Mandarin orange trees planted together with Morishima acacia nine years ago.

3. A peach tree growing in the clover-covered orchard.

4. Thomas, a visitor from The Netherlands, with my grandson.

5. On location in the citrus orchard for a program on national television.

6. Is this confusion or harmonious union? Is it a wood, an orchard, or a vegetable garden?

7. Paradise on Earth? or flight from reality?

8. Vegetables in full bloom beneath a peach tree.

Stopping the Advance of the Desert—A Preliminary Test

1. Sowing seed in a Somalian desert.

2. Ethiopian refugees in Somalia. A natural farm has been set up nearby.

3. The spread of these grasses are the start of the desertification process.

4. (a) Vegetable seeds being sown among the desert grasses.

5. (b) To stop the encroaching desert, first the undesirable grasses are awakened from their dormant state.

6. (c) Preparations are now complete for revegetating the land.

7. (d) After being awakened from dormancy, the desert grasses wither and die.

This forest in California is a living reminder of primeval American forests 2,000 years ago.

These Californian mountains have been turned to semidesert land.

A redwood forest near San Francisco.

Although beautiful, the shore about the Lake of Zurich in Switzerland is an example of nature perverted by livestock grazing and monoculture.

The establishment of a natural farm in Italy.

A farming landscape in Switzerland.

Azienda Farm near Turin, Italy, a large 7,500-acre farm, is practicing natural farming on a grand scale.

This large, 7,500-acre rice farm in California has switched over to natural farming.

The author explains natural farming to farmers on the outskirts of Venice.

The Nelissen natural farm in The Netherlands. Here, apple and pear trees in their natural forms are being grown in a cover of green manure.

In 5 to 20 years, natural farming turns hills of red clay into rich, fertile land. Here, on the author's farm in Shikoku, *daikon* and Indian mustard are in full bloom below peach trees growing about a 6-year-old Morishima acacia.

A natural farm is at once a forest, an orchard, and a vegetable garden. Cherry, peach, plum, wax myrtle, acacia, and green manure plants all bloom together.

The Practice
of Natural Farming

1. Starting a Natural Farm

Once the decision has been made to start farming the natural way the very first problem that comes up is where and on what type of land to live.

Although some may share the woodsman's preference for the isolation and solitude of a mountain forest, the best course generally is to set up a farm at the foot of a hill or mountain. Weather is often most pleasant when the site is slightly elevated. Abundant firewood, vegetables, and other necessities are to be had here, providing all the materials required for food, clothing, and shelter. Having a stream nearby helps make crops easy to grow. This type of location thus provides all the conditions essential for setting up an easy and comfortable life.

Of course, with effort, crops can be made to grow on any type of land, but nothing compares with richly endowed land. The ideal location is one where enormous trees tower above the earth, the soil is deep and a rich black or brown in color, and the water is clear. Scenic beauty perfects the site. A good environment in an attractive setting provides the physical and spiritual elements necessary for living a pleasant life.

The natural farm must be able to supply all the materials and resources essential for food, clothing, and shelter. In addition to fields for growing crops, a complete natural farm should include also a bordering wood.

Keep a Natural Protected Wood

The woods surrounding a natural farm should be treated as a natural preserve for the farm and used as a direct or indirect source of organic fertilizer. The basic strategy for achieving long-term, totally fertilizer-free cultivation on a natural farm is to create deep, fertile soil. There are several ways of doing this. Here are some examples.

1. Direct burial of coarse organic matter deep in the ground.
2. Gradual soil improvement by planting grasses and trees that send roots deep into the soil.
3. Enrichment of the farm by carrying nutrients built up in the humus of the upland woods or forest downhill with rainwater or by other means.

Whatever the means employed, the natural farmer must secure a nearby supply of humus that can serve as a source of soil fertility.

When there is no uphill wood available for use as a preserve, one can always develop a new wood or bamboo grove for this purpose. Although the main function of a preserve is to serve as a deeply verdant natural wood, one should also companion plant trees that enrich the soil, timber trees, trees that supply food for birds and animals, and trees that provide a habitat for the natural enemies of insect pests.

148

Fig. 4.1 Layout of a natural farm on sloping land.

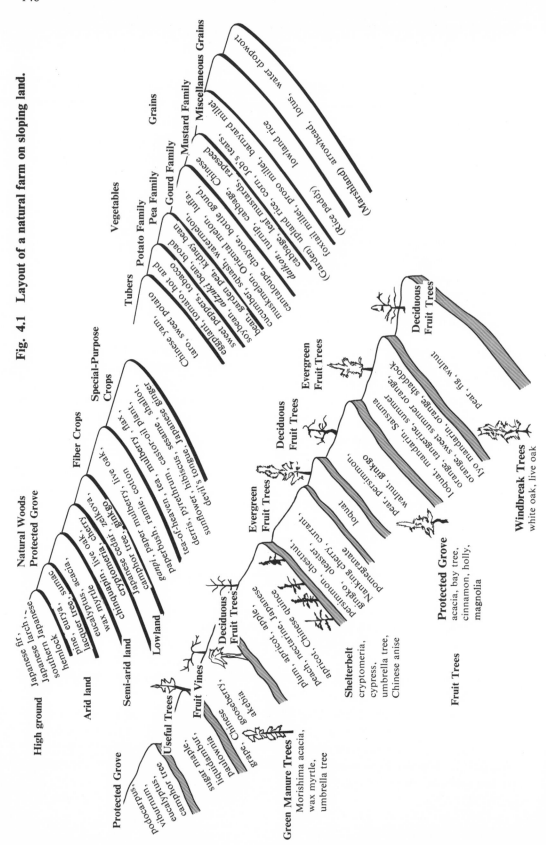

Growing a Wood Preserve: Being generally infertile and dry, hill and mountain tops are highly susceptible to denudation. The first thing to do is plant a vine such as *kudzu* to prevent the soil from washing away. Next, sow the seeds of a low conifer such as moss cypress to create a mountain cover of evergreens. Grasses such as cogon, ferns such as bracken, and low bushes such as lespedeza, eurya, and moss cypress grow thickly at first, but this vegetation gradually gives way to *urajiro* (a fern), *kudzu*, and a mix of trees which further enriches the soil.

Evergreens such as Japanese cypress and the camphor tree should be planted on hillsides, and together with these, deciduous trees such as Chinese hackberry, zelkova, paulownia, cherry, maple, and eucalyptus. Plant the fertile land at the foot of hills and in valleys with oak and evergreens such as cryptomeria and live oak, interplanting these with walnut and ginkgo.

A bamboo grove may serve equally well as the reserve. It takes a bamboo shoot only one year to grow to full size, so the amount of vegetative growth is greater than for ordinary trees. Bamboo is therefore valuable as a source of coarse organic material that can be buried in the ground for soil improvement.

Not only can the shoots of certain species of bamboo be sold as a vegetable, when dried the wood is light and easy to carry. Bamboo is hollow and so has a large void ratio, in addition to which it decomposes slowly. These properties help it to retain air and water in the soil when it is buried. Clearly then, this plant may be used to great advantage in the improvement of soil structure.

Shelterbelts: Shelterbelts and windbreak trees are valuable not only for preventing wind damage, but also for maintaining soil fertility and for environmental improvement.

Fast-growing trees that are commonly planted for this purpose include cedar, cypress, acacia, and the camphor tree. Other species that grow somewhat more slowly but are also used quite often include camellia, the umbrella tree, wax myrtle, and Chinese anise. In some places, evergreen oaks, holly, and other trees are also used.

Setting Up an Orchard

One may establish an orchard and plant nursery stock using essentially the same methods as when planting forest trees. Vegetation on the hillside is cut in lateral strips, and the large trunks, branches, and leaves of the felled trees are arranged or buried in trenches running along hill contours, covered with earth, and allowed to decompose naturally. None of the vegetation cut down in the orchard should be carried away.

A natural farm should be developed without clearing of the land. When land is cleared with a bulldozer, irregular surface features on a slope are flattened and smoothed, allowing wide farm roads to be built that permit farm mechanization. In modern orchards as well, using bulldozers to clear land has become the rule rather than the exception.

However, mechanization really only facilitates certain farm operations such as fertilizer and pesticide application. Since picking ripened fruit is the only major operation in natural farming, there is no need to go out of one's way to clear steep slopes.

Another factor that improves the enterprising orchardist's chances of success is that a natural orchard can be established without a heavy initial outlay of capital or incurring large debts.

Table 4.1 Orchard vegetation.

	Type	Season	Undergrowth
Protected trees **Green manure trees** **Useful trees**	acacia wax myrtle umbrella tree Japanese alder sugar maple bay tree cinnamon	year-round	green manure, vegetables
Green manure crops	ladino clover alfalfa	year-round	
	bur clover	spring	
	Mustard Family vegetables	winter	
	lupine		
	hairy vetch common vetch, Saatwicke	winter	
	soybean, peanut adzuki bean mung bean, cowpea	summer	
Evergreen fruit trees	citrus trees, loquat		butterbur, Japanese silverleaf, buckwheat
Deciduous fruit trees	persimmon, walnut, peach, plum, apricot, pear, apple, cherry		devil's tongue, lily, ginger, buckwheat
Fruit vines	grape, Chinese gooseberry, akebia		barnyard millet, proso millet, foxtail millet

Starting a Garden

People usually think of a garden as a plot of land devoted to the production of vegetables and field crops. However, using the open space in an orchard to raise an undergrowth of special-purpose crops and vegetables is the very picture of nature. Nothing stops the farmer from having his orchard double as a vegetable and grain patch.

Clearly, of course, the system of cultivation and the nature of the garden or orchard will differ significantly depending on whether the principal aim is to grow fruit trees or vegetable crops.

Table 4.2 Base the selection of vegetables to be planted on weed succession. As the garden or orchard matures, a transition will take place in the weeds growing there. Observe the types of weeds growing and plant vegetables belonging to the same family of plants.

Group (Family)	Weeds	Crops
Ferns	*urajiro*, *koshida*, bracken	
Grass Family	eulalia, cogon, foxtail, crabgrass	barnyard millet, foxtail millet, proso millet, wheat, barley, rice
Arum Family	jack-in-the-pulpit	devil's tongue, taro
Yam Family	yam	Chinese yam
Buckwheat Family	knotgrass, knotweed	pigweed, buckwheat, spinach
Composite Family	fleabane, dandelion, thistle, mugwort, aster	garland chrysanthemum, lettuce, burdock
Lily Family	dogtooth violet, gold-banded lily, tulip, asparagus	leek, garlic, shallot, Welsh onion, onion
Mint Family	*hikiokoshi*	perilla, mint, sesame
Pea Family	*kudzu*, common vetch, bur clover, clover	soybean, adzuki bean, kidney bean, garden pea, broad bean
Morning-Glory Family	morning glory	sweet potato
Carrot Family	water hemlock	dropwort, honewort, carrot, parsley, celery
Mustard Family	shepherd's purse	*daikon*, turnip, Chinese cabbage, leaf mustard, cabbage
Gourd Family	snake gourd, bottle gourd	chayote, squash, muskmelon, watermelon, cucumber
Potato Family	ground cherry, sweet brier	hot red pepper, potato, tobacco, eggplant, tomato

Land to be used for growing fruit trees and intercropped with grains or vegetables is prepared in essentially the same way as an orchard. The land does not need to be cleared and leveled, but should be carefully readied by, for example, burying coarse organic material in the ground.

When starting an orchard, the main goals initially should be prevention of weed emergence and maturation of the soil. These can be accomplished by growing buckwheat during the first summer, and sowing rapeseed and Indian mustard that same winter. The following summer, one may plant adzuki bean and mung bean, and in the winter, hairy vetch and other hardy leguminous plants that grow well without fertilizers. The only problem with these is that they tend to inundate the young fruit tree saplings.

As the garden matures, it will support any type of crop.

The Non-Integrated Garden: Gardens are normally created on hillsides and well-drained fields at the foot of larger mountains. Most of the crops grown in these gardens are annuals and the period of cultivation is generally short, in most cases lasting from several months to about a half-year.

Most vegetables rise to a height of no more than three feet or so and are shallow-rooting. The short growing period allows this cycle to be repeated several times a year, providing the surface of the soil with considerable exposure to the sun.

A dry-farmed field, then, is prone to erosion and soil depletion by rainfall, susceptible to drought, and has low resistance to the cold.

Soil movement being the greatest concern when establishing a garden, the garden should be built in terrace fashion with the field surface on each terrace level. The first task in setting up a garden is to build a series of lateral embankments or stone walls running across the slope of the hill.

Knowledge of the soil and the ability to build earthen embankments that do not crumble or to skillfully lay stones dug up from the field can be a determining factor in the success of a garden.

Whether the individual terraces in a terraced garden are level or slightly graded makes a large difference in crop returns and the efficiency of farming work. As I mentioned earlier, the most basic method for improving soil is to bury coarse organic matter in deep trenches. Another good method is to pile soil up to create high ridges. This can be done using the soil brought up while digging contour trenches with a shovel. The dirt should be piled around coarse organic material. Better aeration allows soil in a pile of this sort to mature more quickly than soil in a trench. Such methods soon activate the latent fertility of even depleted, granular soil, rapidly preparing it for fertilizer-free cultivation.

Creating a Rice Paddy

Today, a rice field can easily be prepared by clearing the land with large machinery, removing rocks and stones, and leveling the surface of the field. Yet, although well-suited to increasing the size of single paddy fields and promoting mechanized rice production, such a process is not without its drawbacks:

1) Because it is crude, it leaves a thickness of topsoil that varies with the depth of the bedrock, resulting in uneven areas of crop growth.

2) The load that heavy machinery places on the soil results in excessive settling, causing ground water to stagnate. This situation can induce root rotting and at least partial suppression of initial crop growth on the new field.

3) Levees and walkways are all made of concrete, upsetting and destroying the community of soil microbes. The danger here is of gradually turning the soil into a dead mineral matter.

Traditional Paddy Preparation: Most people might expect open, level ground to be the most sensible place on which to set up rice paddies. But rather than settling on the flat and fertile banks of large rivers, Japanese farmers of old chose to live in mountain valleys where there was far less cause to fear violent flooding and strong winds. They set up small fields in the valleys or built terraced rice fields on the hillsides.

To these farmers, the work of digging channels for drawing water from the valley streams, of constructing rice fields, and of building rock walls and terraced fields was not as hard as the people of today imagine. They did not think of it as hardship.

By spreading the field with the cuttings from ridge grasses, bordering weeds, and young foliage from trees, rice could easily be grown each year without using fertilizers. A tiny field of maybe a hundred square yards supplied the food needs for one individual indefinitely. The spiritual peace and security, the simple joy of creating a rice paddy were beyond what can be imagined. From these activities, our farming ancestors gained pleasure and satisfaction of a sort that cannot be had through mechanized farming.

I can recall occasionally happening upon small paddy fields deep in the mountains far from populated areas and my surprise at how well someone had managed to set up a field in such a location. To the modern economist, this may appear as utter wretchedness, but I found the field a wonderful masterpiece reminiscent of the past, built alone by someone living happily in the seclusion and quiet solitude of the wilds with nature as his sole companion.

In truth, such a place, with its artfully built conduit—snaking in the shade of valley trees—for drawing water, the rockwork that displays a thorough knowledge of the soil and terrain, and the beauty of the moss on the stones, is in reality a splendid garden built with great care by an anonymous farmer close to nature who drew fully on the resources about him.

As the agrarian scenes of yesterday are rapidly swept off by the waves of modernization, we might do well to consider whether we can afford to lose the aesthetic spirit of our farming forbears, who saw the rice paddy as the arbor of their souls and gazed upon a thousand moons reflected in a thousand paddies. But of one thing I am certain: fields and rice paddies imbued with this spirit will reappear again somewhere, someday.

These are not just the fond recollections of bygone days by a misty-eyed old fogey. The general method of establishing a rice paddy I have described here accords with reality as it exists on uncultivated open plains and meadows.

Crop Rotation

Modern farming has brought about destruction of the soil and a loss in soil fertility because it breaks crops up into many different use categories and grows each in isolation, often single-cropping continuously over extensive areas.

On the complete natural farm, fruit trees, vegetables, grains, and other crops must all be planted and grown in an organic and mutually favorable arrangement. More specifically, a reliable crop rotation scheme must be established in order to be able to make essentially permanent use of the land while maintaining soil fertility.

Fruit trees must not be dissociated from the trees of a bordering wood or the weed undergrowth. Indeed, it is only by having intimate associations with these that they are able to show normal, healthy growth. As for vegetables, when left to themselves in a field, they appear at first glance to grow without order, but these develop into splendid plants while nature solves the problems of continuous cropping, space, disease and pest damage, and the recovery of soil fertility.

Ever since primitive man began slash-and-burn agriculture, the question of what crops to plant when has been the greatest problem faced by farmers everywhere, yet a clearly decisive system of crop rotation has yet to be established. In the West, systems of rotation based on pasturage have been established for some time, but because these were designed for the benefit of ranchers and their animals rather than for the land itself, they have brought about a decline in soil fertility that calls for immediate improvement.

In Japan as well, although farmers do grow a wide variety of different crops using an excellent system of crop rotation, a basic crop rotation scheme worthy of more widespread use has yet to be developed. One reason for this is the staggering number of possible crop combinations, and the essentially infinite number of elements that must be considered in stabilizing and increasing yields. To bring all these together into a single system of crop rotation would be an exceedingly difficult undertaking.

The diagrams on the following pages are intended to serve as aids to an understanding of crop rotation.

Rice/Barley Cropping: Japanese farmers have long practiced the continuous rotation of rice with barley. This has enabled them to reap the same harvest year after year indefinitely, something which they have always regarded as perfectly natural. Yet this type of rotational cropping is an extraordinary method of farming that has taken hold nowhere else in the world.

The reason rice and barley can be grown in continuous succession each year is that the rice is grown in paddy fields and soil fertility has been built up by a superior method of irrigation. To tell the truth, I am proud of the outstanding cultivation methods developed by Japanese farmers and would like to see these introduced abroad.

Still, some very simple yet significant improvements could be made. For example, about seventy percent of the nitrogenous components absorbed by rice and barley are supplied directly by the soil, and about thirty percent are furnished artificially by fertilization. If all the straw and chaff from the threshed grain were returned to the fields, farmers would only have to apply at most fifteen percent of the nitrogenous components required by the plants.

Reports have begun appearing recently in scientific journals on the possibilities of developing cultivars of rice not requiring fertilization. These propose the creation of strains of rice capable of fixing nitrogen by incorporating the root nodule genes of soybeans into rice genes. One has to admit, though, that nature has achieved a smarter method of non-fertilizer cultivation. True, because my method of rice-barley cropping under a cover of green manure is, in a sense, just a mimicry of nature, it is incomplete in itself. But there remains much that man can and should try before he resorts to genetic engineering, a technology with the frightening potential to utterly destroy nature.

Upland Rice: Wheat and rice are each the staple foods of about half the world's

Fig. 4.2 Natural continuous cropping system.*

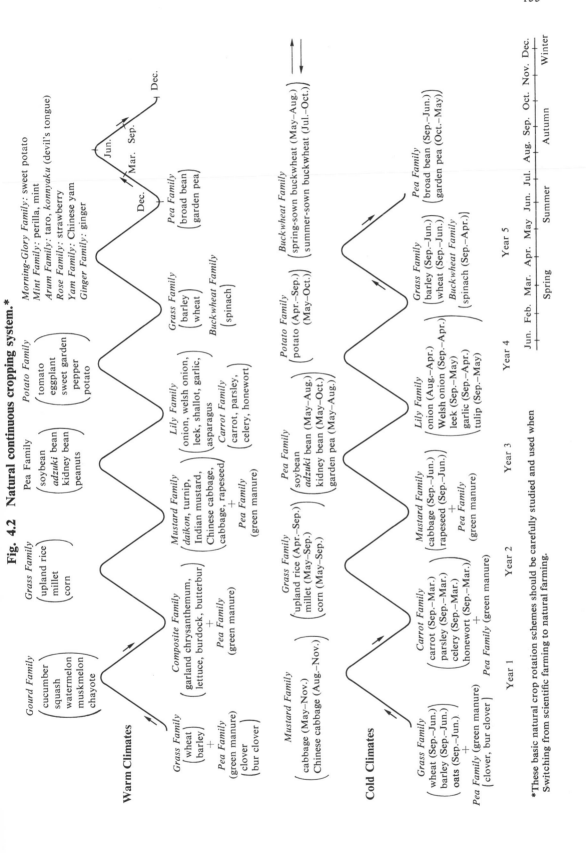

Morning-Glory Family: sweet potato
Mint Family: perilla, mint
Arum Family: taro, konnyaku (devil's tongue)
Rose Family: strawberry
Yam Family: Chinese yam
Ginger Family: ginger

Warm Climates

Grass Family
(wheat
barley)
+
Pea Family
(green manure)
(clover
bur clover)

Gourd Family
(cucumber
squash
watermelon
muskmelon
chayote)

Grass Family
(upland rice
millet
corn)

Pea Family
(soybean
adzuki bean
kidney bean
peanuts)

Potato Family
(tomato
eggplant
sweet garden
pepper
potato)

Composite Family
(garland chrysanthemum,
lettuce, burdock, butterbur)
+
Pea Family
(green manure)

Mustard Family
(daikon, turnip,
Indian mustard,
Chinese cabbage,
cabbage, rapeseed)
+
Pea Family
(green manure)

Lily Family
(onion, welsh onion,
leek, shallot, garlic,
asparagus)
Carrot Family
(carrot, parsley,
celery, honewort)

Grass Family
(barley
wheat)
Buckwheat Family
(spinach)

Pea Family
(broad bean
garden pea)

Dec.
Jun.
Mar. Sep.
Dec.

Cold Climates

Mustard Family
(cabbage (May–Nov.))
(Chinese cabbage (Aug.–Nov.))

Pea Family (green manure)
(clover, bur clover)

Grass Family
(wheat (Sep.–Jun.))
(barley (Sep.–Jun.))
(oats (Sep.–Jun.))
+
Pea Family (green manure)

Carrot Family
(carrot (Sep.–Mar.))
(parsley (Sep.–Mar.))
(celery (Sep.–Mar.))
(honewort (Sep.–Mar.))
+
Pea Family (green manure)

Grass Family
(upland rice (Apr.–Sep.))
(millet (May–Sep.))
(corn (May–Sep.))

Mustard Family
(cabbage (Sep.–Jun.))
(rapeseed (Sep.–Jun.))
+
Pea Family
(green manure)

Pea Family
(soybean
adzuki bean (May–Aug.)
kidney bean (May–Oct.)
garden pea (May–Aug.))

Lily Family
(onion (Aug.–Apr.))
(Welsh onion (Sep.–Apr.))
(leek (Sep.–May))
(garlic (Sep.–Apr.))
(tulip (Sep.–May))

Potato Family
(potato (Apr.–Sep.))
(May–Oct.)

Grass Family
(barley (Sep.–Jun.))
(wheat (Sep.–Jun.))
Buckwheat Family
(spinach (Sep.–Apr.))

Buckwheat Family
(spring-sown buckwheat (May–Aug.))
(summer-sown buckwheat (Jul.–Oct.))

Pea Family
(broad bean (Sep.–Jun.))
(garden pea (Oct.–May))

Year 1   Year 2   Year 3   Year 4   Year 5

| Jun. Feb. Mar. Apr. May Jun. Jul. Aug. Sep. Oct. Nov. Dec. |
| Spring | Summer | Autumn | Winter |

*These basic natural crop rotation schemes should be carefully studied and used when
Switching from scientific farming to natural farming.

156

Fig. 4.3 Crop rotations for major grains and vegetables.*

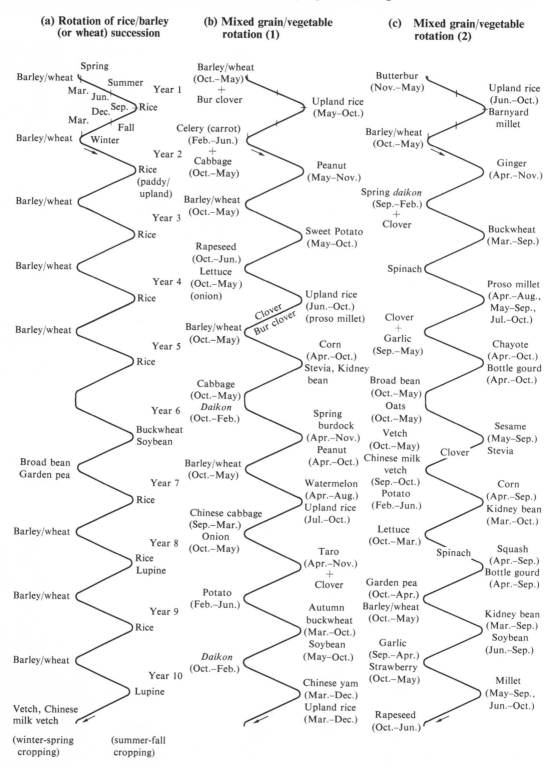

(a) Rotation of rice/barley (or wheat) succession

Spring
Barley/wheat
Mar. Summer
Jun. Year 1
Dec. Sep. Rice
Mar.
Fall
Barley/wheat Winter
Year 2
Rice
(paddy/
upland)
Barley/wheat
Year 3
Rice
Barley/wheat
Year 4
Rice
Barley/wheat
Year 5
Rice
Year 6
Buckwheat
Soybean
Broad bean
Garden pea
Year 7
Rice
Barley/wheat
Year 8
Rice
Lupine
Barley/wheat
Year 9
Rice
Barley/wheat
Year 10
Lupine
Vetch, Chinese
milk vetch

(winter-spring
cropping)
(summer-fall
cropping)

(b) Mixed grain/vegetable rotation (1)

Barley/wheat
(Oct.–May)
+
Bur clover
Upland rice
(May–Oct.)
Celery (carrot)
(Feb.–Jun.)
+
Cabbage
(Oct.–May)
Peanut
(May–Nov.)
Barley/wheat
(Oct.–May)
Sweet Potato
(May–Oct.)
Rapeseed
(Oct.–Jun.)
Lettuce
(Oct.–May)
(onion)
Upland rice
(Jun.–Oct.)
(proso millet)
Clover
Bur clover
Barley/wheat
(Oct.–May)
Corn
(Apr.–Oct.)
Stevia, Kidney
bean
Cabbage
(Oct.–May)
Daikon
(Oct.–Feb.)
Spring
burdock
(Apr.–Nov.)
Peanut
(Apr.–Oct.)
Barley/wheat
(Oct.–May)
Watermelon
(Apr.–Aug.)
Upland rice
(Jul.–Oct.)
Chinese cabbage
(Sep.–Mar.)
Onion
(Oct.–May)
Taro
(Apr.–Nov.)
+
Clover
Potato
(Feb.–Jun.)
Autumn
buckwheat
(Mar.–Oct.)
Soybean
(May–Oct.)
Daikon
(Oct.–Feb.)
Chinese yam
(Mar.–Dec.)
Upland rice
(Mar.–Dec.)

(c) Mixed grain/vegetable rotation (2)

Butterbur
(Nov.–May)
Upland rice
(Jun.–Oct.)
Barnyard
millet
Barley/wheat
(Oct.–May)
Ginger
(Apr.–Nov.)
Spring daikon
(Sep.–Feb.)
+
Clover
Buckwheat
(Mar.–Sep.)
Spinach
Proso millet
(Apr.–Aug.,
May–Sep.,
Jul.–Oct.)
Clover
+
Garlic
(Sep.–May)
Chayote
(Apr.–Oct.)
Bottle gourd
(Apr.–Oct.)
Broad bean
(Oct.–May)
Oats
(Oct.–May)
Vetch
(Oct.–May)
Sesame
(May–Sep.)
Stevia
Chinese milk
vetch
(Sep.–Oct.)
Clover
Potato
(Feb.–Jun.)
Corn
(Apr.–Sep.)
Kidney bean
(Mar.–Oct.)
Lettuce
(Oct.–Mar.)
Spinach
Squash
(Apr.–Sep.)
Bottle gourd
(Apr.–Sep.)
Garden pea
(Oct.–Apr.)
Barley/wheat
(Oct.–May)
Kidney bean
(Mar.–Sep.)
Garlic
(Sep.–Apr.)
Soybean
(Jun.–Sep.)
Strawberry
(Oct.–May)
Millet
(May–Sep.,
Jun.–Oct.)
Rapeseed
(Oct.–Jun.)

157

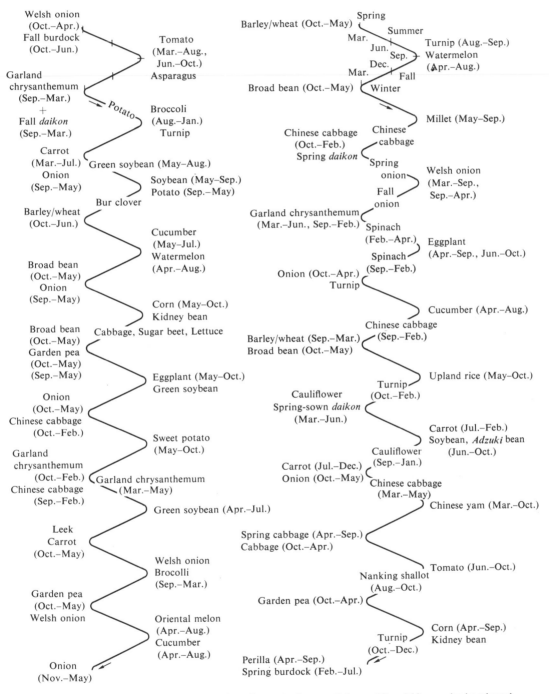

(d) Vegetable rotation (1)

(e) Vegetable rotation (2)

*Rotation schemes (a–c) are for use by farmers. Schemes (d) and (e) may also be adapted for family vegetable gardens.

population, but if the cultivation of upland rice were to spread and this grain became easy to harvest in high yield, a large jump would occur in the number of rice-eating peoples. Growing upland rice could even possibly become one effective way of coping with the worldwide scarcity of food.

Generally speaking, upland rice is an unstable crop often subject to drought. Yields are lower than for rice grown in paddy fields, and continuous cropping gradually depletes soil fertility, resulting in a steady decline in yields. A workable solution appears to be rotational cropping in combination with various green manure crops and vegetables, as this raises the ability of the soil to retain water and gradually builds up soil fertility.

Minor Grains: This group includes members of the grass family such as millet and corn, as well as buckwheat, Job's tears, and other grains. Compared with rice, barley, and wheat, these grains generally receive short shrift because of their "inferior" taste and a lack of research on methods for their use, but they deserve more attention for their very great value as prototypic health foods essential for maintaining the physical well-being of human beings.

The same is true also for vegetables and other plants in general. The wilder and more primitive the food, the greater its medicinal value.

With changes in popular taste, the cultivation of these minor grains as food for man has rapidly receded to the point where even seed preservation has become difficult. Yet, above and beyond their importance as a food for humans and animals, they have also played a vital role as coarse organic matter essential for soil preservation.

When single-cropped or grown continuously, these grains deplete the soil, but if rotated with green manure crops and root vegetables, they improve and enrich the soil. This is why I believe the minor grains should be repopularized.

Vegetables: People tend to think of vegetables as frail crops that are difficult to grow, but with the exception of several types that have been genetically over-improved, such as the cucumber and tomato, these are surprisingly hardy crops that can thrive even under extensive cultivation.

Cruciferous winter vegetables, for example, when sown just before the emergence of weeds, grow vigorously, overwhelming the weeds. These also send down roots deep into the soil, and so are highly effective in soil improvement. That leguminous green manure suppresses summer weeds and enriches the soil hardly needs repeating. Clearly these too should play an important part in a crop rotation.

Judicious combinations of vegetables in a sensible mixed cropping scheme can be grown in fair yield, free of disease and pest damage, without resorting to pesticides. I have found also, through personal experience, that most vegetables, when cultivated in a semi-wild state that could be considered a natural rotation, can be grown almost entirely without fertilizers.

Fruit Trees and Crop Rotation: Because fruit trees are continuously cultivated perennials, they are subject to the difficulties associated with continuous cropping.

The purpose of having a protected wood and a ground cover of weeds is to resolve such problems naturally and extend the life of the fruit trees. These trees exist, together with the companion-planted manure trees and the weed under-growth, in a three-dimensional rotational cropping relationship.

When vegetables are grown beneath fruit trees, the number of insect pests tends to be low. Some diseases and pests are common both to fruit trees and vegetables, and some are not. These in turn have a host of different natural enemies that emerge at various times of the year. As long as a balance is maintained between the fruit trees, the vegetables, the insect pests, and their natural predators, real damage from disease and insect attack can be prevented. For the same reason, the planting of manure trees and windbreak trees, and the combination planting of evergreens and deciduous trees may also be helpful in diminishing damage.

In most cases, serious disease and pest damage in fruit trees, such as by long-horned beetles and scale insects, is triggered by diminished tree vigor due to depleted soil fertility, a confused tree shape, poor ventilation, inadequate light penetration, or a combination of all of these factors. Because they help sustain soil fertility, a ground cover of green manure crops and the combination planting of manure trees may thus be regarded as basic defensive measures against disease and pest damage.

Using natural farming methods to cultivate fruit trees creates a truly three-dimensional orchard. More than just a place for growing fruit, the orchard becomes an organically integrated community that includes fowl, livestock, and man as well. If a natural orchard is managed and run as a single microcosm, there is no reason why one should not be able to live self-sufficiently.

By looking with equal detachment at insects, which man categorizes as beneficial or harmful, people will see that this is a world of coexistence and mutual benefit, and will come to understand that farming methods which call for heavy inputs of fertilizer and energy can only succeed in robbing the land of its natural fertility.

Nature is sufficient in and of itself; there never was a need for human effort and knowledge. By returning to a "do-nothing" nature, all problems are resolved.

2. Rice and Winter Grain

The Course of Rice Cultivation in Japan

In the Land of Ripening Grain, as the Japanese people have long been fond of calling their country, rice cultivation held a deeper meaning for farmers than simply the growing of a staple food crop. The farmer did not grow the rice, nature did; and the people born to this land partook of its blessings. The words "bountiful Land of Ripening Grain" expressed the joy of the Yamato people, who were able to receive the rich blessings of heaven and earth with a grateful heart.

However, once man began to think that he grew the rice, scientific discrimination arose, creating a rift between the rice and the land. People lost a sense of unity with nature, leaving in its place only man's relationship with rice cultivation and his relationship with the soil.

Modern thinking reduced rice to just another foodstuff. It began to view the work of farmers engaged in rice cultivation—service to God—as an economically inefficient and unscientific activity. Yet has rice really been just a food, a material object, all along? Was the labor of farmers merely one field of economic activity? And have farmers been nothing more than laborers engaged in food production?

The Japanese people have lost sight of the true value of rice. They have forgotten the spirit of gratitude with which farmers made offerings of their ripened rice to the gods to celebrate the fruits of autumn. From the scientific perspective, this substance we call rice has a value equivalent only to its nutritional value as a human food. Although the ripened grain may be seen as a reward for human labor, there is no joy in the knowledge of this as the product of a common effort by heaven, earth, and man. Nor is there any awe at the emergence of this life of infinite majesty from nature's midst. More than just the staff of life, the rice grown on Japanese soil was the very soul of the Yamato people.

But as the activities of the farmer have been lowered in the common perception to rice production as another foodstuff, a commercial article, the original purpose of rice production has gradually been corrupted. The object no longer is the cultivation of rice, but starch production, and more precisely, the pursuit of profits through the manufacture and sale of starch. A natural consequence of this can be seen in the efforts by farmers today to raise income by raising yields.

Changes in Rice Cultivation Methods: Rice farming in Japan has passed through several stages recently which can be represented as follows:

1) 1940—*Primitive farming* (improvements in tilling methods)
2) 1950—*Animal-powered agriculture* (increased fertilizer production)
3) 1960—*Scientific farming* (mechanization)
4) 1970—*Agribusiness* (energy-intensive systemized agriculture)

Prior to the development of scientific agriculture, rice farmers devoted themselves entirely to serving the land that grows the crops. But they gradually turned their attention from the land to the problem of boosting soil fertility and discussion came to dwell on what constitutes soil fertility.

Those familiar with the recent history of Japanese farming will know that, once it became clear that the most effective way to boost soil fertility was to till more deeply and add more organic material to the soil, campaigns to improve plows and hoes and to increase compost production from grass cuttings and straw spread throughout the country. Soil scientists showed that tilling the soil to a depth of one inch can yield five bushels of rice per quarter-acre, and from this concluded that working the soil down to five inches would yield 25 bushels.

Animal-powered agriculture was later pushed because heavy applications of manure and prepared compost were known to help achieve high yields. Farmers learned, however, that preparing compost is not easy work. Yields failed to improve enough to justify the heavy labor required, peaking at about 22 bushels per quarter-acre. Efforts to push yields even higher resulted in unstable practices, relegating animal-powered agriculture largely to the status of a model practice used by few farmers.

Much research is being done today on the morphology of rice at various stages of growth. Scientists are attempting also to achieve high yields through detailed comparative studies on the planting period, quantity of seed sown, number and spacing of transplanted seedlings, and depth of transplantation. However, because none of the resulting techniques has more than, say, a five percent effect on yields, efforts are underway to combine and consolidate these into one unified high-yielding technology.

Yet such efforts have failed to make any notable gains, save for occasional increases in yield in low-yielding areas through basic improvements, better water drainage, and other correctives. Although Japanese agricultural technology appears to have made rapid progress over the last fifty years, the productivity of the land has declined. In terms of quality, this period has been one of retreat rather than advance.

Because the emphasis in paddy-field rice production today is on the productivity of labor, farmers scramble after returns and profits; they have abandoned animal-powered farming and wholeheartedly embraced scientific farming, especially mechanization and the use of chemicals. Much has been made of the organic farming methods taken up by a small number of farmers out of concern over the polluting effects of scientific farming, but organic farming too is an outgrowth of scientific farming that is oriented towards petroleum energy-intensive commercial agribusiness.

The only course available today for successfully rejecting scientific farming and halting its rampant growth is the establishment of a natural way of farming the agricultural mainstays: rice, barley, and wheat.

Until recently, barley and wheat, grown in most parts of Japan as winter grains, have been second only to rice in their importance as food staples of the Japanese people. Along with brown rice, the taste of cooked rice and barley was something dear to Japanese farmers. Yet these winter grains are today in the process of vanishing from Japanese soil. As recently as fifteen or twenty years ago, the paddy field was not neglected after the rice harvest in the fall; something was always grown there during the winter months. Farmers knew that productivity per unit area of paddy was never better than when a summer rice crop was followed by a crop of barley or wheat in the winter.

As soon as the rice was harvested in the fall, the paddy field was plowed, ridges formed, and the barley or wheat seed sown. This was done because winter grain was thought to have a poor resistance to moisture.

Planting barley was no easy process. The farmer began by plowing up the field. He then broke up the clods of earth, made seed furrows, sowed the seed in the furrows, covered the seed with dirt, and applied prepared compost. When this process was finally over, but before the year was out, he had to do the first weeding. He followed this early in the new year with a second and third weeding. While weeding, he passed his hoe along the rows, loosening the soil. Then he would gather additional earth around the base of the plants to prevent frost damage, and trample the shoots to promote root growth. After repeating this process several times, he sprayed the young plants twice with pesticide and left them to mature. All this work was done during the cold months, but harvesting time came at the end of May, which felt even more swelteringly hot than mid-summer. What's more, if the crop was late-maturing wheat or barley, the harvest usually took place during the rainy season, which meant farmers had to go through the considerable trouble of drying the harvested grain. Winter grain cultivation, then, was a very taxing process.

Some fifty years ago, domestic wheat varieties were improved and the use of wheat encouraged to hold down wheat imports from the United States. Wheat was widely planted in place of barley and naked barley, but wheat grown for bread-making is late-maturing for the Japanese climate and so its use resulted in unstable harvests.

Then, from around 1945, the Japanese Ministry of Agriculture and Forestry, deciding that wheat grown domestically could not compete with cheaper foreign-grown grain, adopted a policy of increased dependence on other countries for the supply of food and feed provisions. This had the effect of causing farmers in the domestic wheat belt to abandon their production of wheat.

It was neither money nor labor that supported the arduous practice of double cropping paddy fields with wheat or barley. It was pride. The farmer, afraid of being called lazy or wasteful if he left his fields fallow over the winter, plowed every inch of available Japanese soil. So when the farming authorities started saying that nobody had any need for expensive wheat and talking about a euthanasia of domestic wheat production, this knocked the moral support out from under the farmer, speeding his physical and spiritual downfall. Over the past five years or so, wheat and barley production has almost disappeared in some localities.

Thirty years ago, Japan was still essentially self-sufficient in food production, but over the last several years, calorie self-sufficiency has dropped below the 40 percent level. This caused many to question Japan's ability to secure necessary food resources and has led once again to encouragement of domestic wheat and barley production. But is it really possible to revive the former pride and spirit of the farmer?

Back when everyone was sold on the idea that domestic wheat production was unnecessary, I kept telling people that there is a method of wheat and barley cropping that will give us grain as inexpensive as foreign grain, that the prices of farm products should basically be the same everywhere, and that the only reason they are not is because economic manipulations have made prices higher for some and lower for others.

Few field crops yield as many calories as barley. This crop is well-suited to the Japanese climate and should be double-cropped, as in the past, with rice. With a little resourceful planning and effort, most Japanese paddy fields could be readied for growing winter grain. Knowing this, I have consistently maintained that a continuous rice and barley or wheat succession must be made the mainstay of Japanese agriculture.

Natural Barley/Wheat Cropping: I passed through three stages in moving toward the natural cultivation of barley and wheat: 1) tillage and ridge cultivation, 2) level-row, light-tillage or no-tillage cultivation, and 3) natural cropping based on no-tillage cultivation.

1. *Tillage, ridging, and drilling:* In Japan, naked barley and wheat seed was normally drilled at a seeding width of 6 to 7 inches on ridges spaced 3 feet apart.

Forty years ago, most farmers and agricultural experts thought that broad, shallow seeding gave high yields, so I tried increasing the sowing area by 25 percent, 30 percent, and 40 percent. First I increased the seeding width to 10 to 12 inches or more; not only was there no observable improvement in yield, this reduced stability of the crop. I then tried sowing in two rows per ridge at a seeding width of 7 to 10 inches in ridges 4 feet apart, but this resulted in excessive vegetative growth and a small number of heads.

Noting that a narrower seeding width increases yield, I reduced the width and increased the distance between rows. By sowing in two rows on ridges spaced 3 feet apart and setting the rows far enough apart to prevent plants in adjacent rows from crowding each other, I was able to raise my yields. But this sowing method made the furrows between ridges narrower and shallower and reduced ridge height, so that all intertilling and weeding had to be done entirely by hoe.

To increase harvest yields, I raised the number of rows per ridge from two to three, then four. Recently, farmers have taken narrow seeding widths a step further and are drilling seeds in single file.

2. *Light-tillage, low-ridge or level-row cultivation:* Since seeding in three or four rows on a 3-foot ridge results in a low ridge almost level with the ground, I switched to light-tillage and drilled individual seeds in straight, narrow rows.

Although I had thought that naked barley had to be grown on high ridges, I found that it can be grown using a simple light-tillage method. I noticed, moreover, that because the young barley shoots are susceptible to moisture damage during light-tillage, a no-tillage process works even better. So in 1950, I began studying seeding techniques that would allow me to drill narrow rows on an unplowed field. This set me on the road toward a natural method of growing barley and wheat.

There remained the problem of weed control, however. I tried sowing ladino clover as a ground cover together with the barley, and scattered rice straw over the planted field. No farmer at the time spread his paddy fields with fresh straw and agricultural experts strictly forbade anyone from leaving straw on the paddy for fear of disease. I went ahead and used rice straw anyway because I had earlier confirmed beyond any doubt that rice straw left on the ground during the autumn decomposes entirely by the following spring, leaving no trace of pathogenic microbes. This cover of fresh straw showed great promise in weed control.

3. *No-tillage, direct-seeding cultivation:* I built an experimental seeding device and tried dibbling, then drilling, and finally individual seeding in furrows. As I was doing this, and making full use of a straw cover, I grew increasingly certain of the validity of direct seeding without tillage.

I went from sparse seeding to dense seeding, then returned again to sparse seeding before I settled on my present method of broadcasting seed.

My experiments convinced me of the following:

a) No-tillage cultivation not only does not degrade the land being worked, it actually improves and enriches it. This was demonstrated by more than ten years of no-tillage direct-seeded rice/winter grain succession cropping.

b) This method of cultivation is extremely simple, yet it provides total germination and weed control, and is less laborious and higher yielding than other methods.

c) The full potential of this method can be tapped only by combining it in a natural farming rotation with direct-seeded rice.

From the very outset, I had wondered why rice and barley, both members of the grass family, should be grown so differently. Why was it that barley could be sown directly while rice had to be seeded in a starter bed then transplanted? And why was it that barley was grown on ridges while rice was grown on a level field?

All along, I had felt that the most natural method of cultivation for both was direct seeding on a level field. Yet, for a long time the idea that rice and barley could be grown in the same way was nothing more than pure conjecture.

But after long years of failure upon failure, somehow my methods of growing rice and barley merged. I found mixed seeding and even simultaneous seeding to be possible. This is when I became convinced that I had at last arrived at the foundation for a natural way of farming.

Table 4.3 Naked barley* yield—1965.

(The Fukuoka Farm)   (survey by Ehime Prefectural Agricultural Testing Center)

| | Yield of milled grain | | Weight per 1,000 grains (oz.) | Grade |
	(lb./1/4-acre)	(oz./yd²)		
Section A	1,450	21.1	0.94	Good
Section B	1,314	21.2	0.91	Very Good

Section A: 8 sample quadrats on 1/4-acre fertilized field
Section B: 8 sample quadrats on 1/4-acre unfertilized field

Actual yield on 1 acre was 5,488 lbs. of milled grain plus 201 lb. of gleanings.

Growth Survey: average tillers per plant   23–32
average heads per plant   1,800–2,500
average grains per head   62–72

*Variety: early-maturing Hinode

Fig. 4.4   Progression of seeding methods.

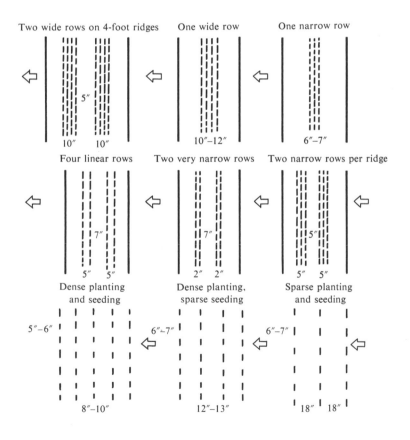

166

Fig. 4.5 Relationship of barley yield to plant growth.

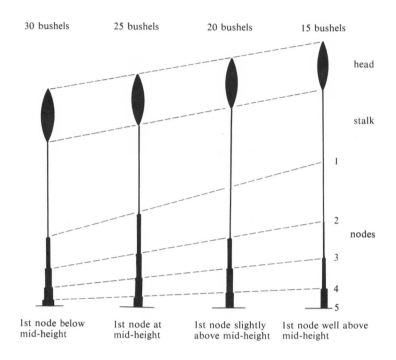

30 bushels   25 bushels   20 bushels   15 bushels

head

stalk

1

2   nodes

3

4

5

1st node below   1st node at   1st node slightly   1st node well above
mid-height   mid-height   above mid-height   mid-height

Early Experiences with Rice Cultivation

As a youth, I set out first to become an agricultural specialist. Being the eldest son in a farming family, I knew that I would have to return to the land someday, but until that time came I was determined to travel a free road.

My field of specialty was plant pathology. I learned the basics from Makoto Hiura at Gifu Agricultural High School and got my practical training under Suehiko Igata of the Okayama Prefecture Agricultural Testing Center. Afterwards, I transferred to the Plant Inspection Division of the Yokohama Customs Bureau, where I did research under Eiichi Kurosawa at the Division's research laboratory in Yamate. I had embarked on a most ordinary course in life and could have spent those early years in the full bliss of youth.

But my fate flew off in an unexpected direction. I had been grappling with the meaning of life and humanity when one night the truth came to me in a flash. I saw all of a sudden that nature is an astounding thing that cannot be named. In that instant, I understood the principle of "nothingness," of Mu. This later gave birth to my method of natural farming, but at first I was totally absorbed by the conviction that "there is nothing in this world. Man should live only in accordance with nature. He has no need to do anything."

Researchers at agricultural testing stations still had a measure of freedom in 1940. I did my work at the plant disease and pest section with just the right measure of diligence and was thus able to live within my dreams. I was fortunate indeed, as a heretic, to have the freedom of working within science and exploring how to refute science and technology.

However, as the war situation intensified, raising food production became a more urgent priority than basic scientific research and so all researchers at the laboratory were mobilized. The directives stated that starch production was to be increased, even if this meant cutting the production of other crops. I was sent to the agricultural testing station in Kochi Prefecture.

While I was there, the local agricultural administration implemented a bold new plan of a type rarely attempted before. This called for the eradication of the yellow rice borer through post-season rice cultivation. Because post-season cultivation made collective use of the most advanced rice growing technology of the day, knowing something about this method gave one a good idea of where scientific farming stood technically at the time.

Rice cultivation practices in Kochi Prefecture were everywhere different. Farmers on the centrally located Kacho Plain, for example, double-cropped their rice, while farmers in other areas of the prefecture variously practiced early-season, midseason, or late-season cropping pretty much as they wished. As a result, transplanting started in April and continued on through to the beginning of August.

In spite of its warm climate, which seemed ideal for rice production, Kochi Prefecture had the lowest rice yields in Japan, save for Kagoshima Prefecture. What was needed here then was not technology for expanding production so much as an understanding of the causes for the low yields. The situation called for the immediate development of methods to stem production losses. I remember commenting on how there wasn't "a single healthy rice plant on the Kacho Plain," an indiscretion for which I was roundly criticized. But facts are facts, and there was no disputing that to increase production in Kochi Prefecture, the first step would have to be the curbing of production losses by diseases and pests. The upshot was that a plan for eradicating the yellow rice borer was drawn up, leading to the promulgation by prefectural edict of a rice cultivation control ordinance.

All the scientists and technicians in the prefectural crop production, agricultural testing, and agricultural cooperatives divisions joined in a common effort to guide the prefecture's farmers in carrying out this post-season cultivation program.

Now that I think on it, although this happened during the war, I cannot help marveling at how such an ambitious pest control program was conducted. Not only was this sort of rice-growing reformation virtually unheard-of in Kochi Prefecture, it is rare in the annals of rice cultivation in Japan. The program was to be carried out in phases, covering a different part of the prefecture during each of three successive years.

We took advantage of the fact that the yellow rice borer does not feed on plants other than rice. The idea was to eliminate the rice borers through starvation by ensuring the absence of all rice plants during the first period of rice borer emergence. Farmers in a one- or two-district area were forbidden from planting rice until the 8th of July (July 3rd the second year). Although the reasoning behind this eradication plan was extremely simple, I can remember agonizing over which day in July to set as the end of the rice borer's first emergence period. A mistake would have been a very serious matter.

Specialists in another area had it even tougher. Waiting until early July to begin growing rice meant drastically shortening the growing season, a risky proposition for both the farmer and the technician. This was Kochi, where farmers began transplanting very early-season rice in April and continued planting early-season, midseason, and late-season rice, followed in some cases by a second crop, right through to early August. Add also the fact that local farmers saw this as the best possible method of cultivation in their area, both in terms of business and improving yields. It should not be hard to imagine then how much trouble we had in gaining the understanding and cooperation of farmers with a program that brought local growing practices under government control and placed all bets on a single post-season rice crop that could not be transplanted until early July.

Other technicians had their hands full too, since all tilling and seeding methods, as well as the fertilizing schedules, had to be changed to accord with July transplanting. There were also many other changes to make, such as modifications in cultivation practices and in the rice cultivars used. It was a true technical reformation in every respect.

The Crop Science Division, for example, had to take measures to cope with delayed transplantion. These included 1) increasing the number of rice plants and seedlings transplanted to the paddy; 2) expanding the size of nursery beds; 3) getting farmers to prepare raised, semi-irrigated rice seed beds; 4) selection of post-season varieties and procurement of seed rice; 5) securing labor and materials; and 6) overseeing the preceding barley crop. The Fertilizer Division had its hands full with changes in the fertilization schedule and making sure that farmers adhered to the new schedule. They had to come up with a schedule that would curb declines in harvests from post-season cultivation and actually push for expanded production. Specialists in each division were expected to be familiar with plans and affairs in all other divisions. Professional opinions from each division were combined into a single collective plan of action. All specialists acted in concert, and familiarizing themselves with the same overall set of techniques in the program, went out one by one to their appointed towns and villages where they supervised local implementation of the program.

Before the prefectural edict was issued, local farmers lodged a hundred objections against post-season rice cultivation, but once the policy was set, the farmers of Kochi Prefecture made a full about-face and gave their total, undivided cooperation. It was an enterprise carried out on a grand scale.

Second Thoughts on Post-Season Rice Cultivation

The outcome of the Kochi Prefecture post-season cultivation program, conducted to exterminate the yellow rice borer and increase food production through rice/barley double-cropping, was mixed: the yellow rice borer was completely eliminated, but we were unable to increase crop production. What is one to make of these results?

First, it might be good to examine the viability of post-season cultivation as a means for controlling the rice borer. Just how well was the real extent of rice borer damage investigated and understood initially? Damage by rice borers always tends to be overestimated since white heads of grain due to post-heading damage stand out in the field. This degree of damage is often mistakenly assumed to translate directly into harvest losses. Even when the crop seems a total loss, damage is generally at most about thirty percent and actual losses are no more than twenty percent. Even during severe infestations, damage is generally at most ten to twenty percent. More importantly, the reduction in final yield is almost always under ten percent, and often even less than five percent. The overall rate of damage over a wide area then is usually grossly overestimated.

Damage by disease and insect pests is usually highly localized. Even in a large, regional outbreak of rice borer, close examination reveals widely differing degrees of infestation; there may be some fields with thirty percent damage and others with virtually no damage at all. Science prefers to overlook those fields that have been spared and focus instead on severely infested fields. (Natural farming, on the other hand, devotes its attention to the fields that escape damage.)

If one small section of a large rice field contains rice grown with lots of fertilizer, rice borers congregate on this soft, vulnerable rice. The farmer could take advantage of this behavior by collecting the insects in one area and destroying them, but what would happen if he left them alone? Although one might expect them to spread out to surrounding fields and cause extensive damage, this just isn't so. Damage would be limited to the small sacrificial area—maybe no more than one percent of the field under cultivation.

During the fall, sparrows gather about the ripening heads of grain, causing serious damage. If, unable to stand by and do nothing, one puts out scarecrows to chase the birds away, then the farmer in the next field feels he has to put scarecrows out too. This snowballs, and before you know it, everyone in the village is busy chasing away sparrows and laying mist nets over their fields to keep the birds out. Does this mean that if no one did anything the sparrows would devastate the fields? Certainly not. The number of sparrows is not determined simply by the amount of grain available. Other factors such as minor crops and the presence of bamboo groves in which to roost all come into play. So do climatic factors such as snow in the winter and summer heat, and of course, natural enemies. Sparrows do not multiply suddenly when the rice begins heading.

The same is true also of rice borers. They do not multiply or go into a decline all of a sudden simply because of the amount of rice growing. Rice borers were singled out in Kochi because they feed only on rice. Nature does not go on unbalanced rampages. It has mechanisms for self-control in places unknown to man.

What sense does it make if, having exterminated yellow rice borers, damage by rice stem borers and cutworms increases? Insect pests and crop diseases sometimes offset each other. On the other hand, a decline in insect infestation, followed by rice blast disease or sclerotium rot can open up a new can of worms. No in-depth study was conducted, so there is no way of knowing for certain, but the lack of a significant increase in yields despite elimination of the rice borers suggests that this is what may have happened in Kochi.

The first thing that pops into the head of an agricultural scientist when he sees a pest emerge in the fields is how to kill it. Instead, he should examine the causes of the outbreak and cut off the problem at its roots. This, at any rate, is the way natural farming would handle the matter. Of course, scientific farming does not neglect, in its own way, to determine the cause of rice borer emergence and take measures against this. It was easy enough at Kochi to imagine that the large infestation of yellow rice borers probably arose from developments in vegetable growing such as the spread of forced vegetable cultivation. This and other factors, including the disorderly and continuous planting of rice, provided an ideal environment for just such an outbreak.

But, we doubled back before finding the true cause and concentrated all our efforts on eradication of the visible pest. For instance, we did not bother to investigate whether the disorder in the rice planting schedules invites outbreaks of the rice borer. The number of borers that emerge in the first generation each year is thought to be dependent on normal overwintering of the insects, but so long as the connection between the rice stubble in which the borer spends the winter and the chaotic local planting practices remains unclear, one cannot attribute a borer outbreak to disorderly planting merely because lots of food is available for the borers. There must have been other reasons why the yellow rice borer, rice stem borer, and other insect pests were so numerous in Kochi Prefecture. I think that the cause had less to do with the environment than with poor methods of cultivating rice.

There is something basically wrong with arbitrarily deciding that this insect in front of one is a pest and trying to destroy it. Before the war, attempts were made to wipe out the rice borer by putting up light traps all over the Kochi Plain. The same thing was tried again after the war with a blanket application of organophosphate pesticides. The campaign against the yellow rice borer through post-season cultivation may have appeared as an indirect and drastic measure, but eradicating one pest out of dozens was bound to end up as nothing more than a temporary expedient.

It must be remembered that diseases and pest damage are self-defense measures taken by nature to restore balance when the natural order has been disturbed. Pests are a divine warning that something has gone wrong, that the natural balance of rice plants has been upset. People must realize that nature's way of restoring an abnormal or diseased body is to fight fire with fire, to use naturally occurring disease and infestation to counter further disease and pest damage.

Rice growth in Kochi Prefecture, with its warm temperatures and high humidity, is too luxuriant. Disease and pest attack is one method taken by nature for suppressing excessive growth, but man applies a near-sighted interpretation, seeing such damage rather as injury and harm. These outbreaks have a role to play in the natural scheme of things.

If someone were to ask me then just how successful our post-season cultivation program in Kochi had been at increasing food production—the goal of the program, I would have to answer that such cultivation, in spite of the daring methods used, never had the makings of an enduring yield-increasing technique.

Even in the selection of a cultivar, for example, scientific farming normally chooses a thermosensitive variety for early planting and a photosensitive variety for late planting, so for post-season cultivation we factored in both photosensitivity and cumulative temperature, and selected a cultivar appropriate for July planting. What we were doing, very simply, was selecting a cultivar suited to an artificially chosen period. There were no real standards to guide us. The only role of the cultivar was to meet certain goals established according to the needs of the moment. The post-season cultivar selected was merely one that would not reduce yields when planted in July; in no way was it capable of positively raising yields.

We had no idea either of what the best time was for planting, a factor thought to play a key role in determining yields. We chose post-season planting simply as a measure against the rice borer.

Crop cultivation techniques based on late planting are all merely stopgap measures for holding crop losses to a minimum. These, like the techniques we employed in post-season cultivation, have no other effect than to maintain the status quo.

That this post-season cultivation program, which represented a cross-section of the most advanced agricultural technology of the time, succeeded only in preventing further losses was very significant, for it demonstrated that, since the purpose of scientific agriculture is always and everywhere convenience to man, no matter how large and complete the technology amassed, it will never amount to more than a temporary expedient.

This incident taught me not to rely on human action and strengthened my resolve to move toward a natural way of farming.

First Steps Toward Natural Rice Farming

At Kochi, while I took part in the common effort to scientifically increase food production, I inwardly searched for what I believed to be the true path of agriculture—natural farming. I had yet no clear image of natural farming; all I could do was grope blindly for a way of farming I had never seen but knew must exist. During this period, I did stumble across a number of important clues, one of which was the ability of nature to "plant without sowing seed."

Natural Seeding: The year that we began our program of post-season cultivation to eradicate the yellow rice borer, I was assigned to an eastern district of Kochi Prefecture. My job was to make certain that not a single stalk of rice remained standing as food for the season's first generation of rice borers until the end of June. I combed the entire district, making my rounds from the hilly back country and mountains to the coast.

Once, as I was passing through a pine wood along the shore at Kotogahama, I spotted a large number of young rice seedlings that had sprouted from unhulled seed spilled where farmers had threshed rice the year before. This volunteer rice later led to my method of biennial, or overwintering, cultivation. Curiously enough, having caught sight of this once, I later noticed, again and again, overwintered rice germinating from seed still attached to rice straw.

Nature then "plants without sowing seed." This realization was my first step toward natural rice cropping, but it was not enough in itself. I learned from this only that rice seed sown by man in the autumn does not easily survive the winter.

In nature, the grain ripens in the autumn and falls to the ground as the leaves and stalks of the rice plant wither and die. And yet, nature is very subtle. Long ago, rice shattered as easily as other grasses, the grains falling in a certain order, starting at the top of the panicle and going on down. The chances of a seed that falls to the ground of surviving intact until the following spring are less than one in a million. Almost all are consumed by birds and rodents or destroyed by disease. Nature can be a very cruel world.

However, a closer look reveals that the vast quantity of grain which appears as unnecessary waste serves a very important purpose by providing food for insects and small animals during the winter months. But nature was not so indulgent as to leave enough grain lying around to feed people who sit and do nothing.

Well over ten years later, I finally succeeded in developing a long-lasting protectant—consisting of a mixture of pesticide and synthetic resin—with which to coat rice seed for protection against winter damage by rodents and other pests. My next step was to eliminate the need for this protectant, which I was able to do by sowing seed enclosed in clay pellets.

While at Kochi, I also observed shoots growing from rice stubble in harvested fields. I was traveling all over the prefecture investigating how summer and fall leafhoppers overwinter—of which little was known at the time—when I observed the ability of regenerated rice shoots and certain harmful grasses to survive the winter.

In areas not hit by frost, it should be possible to make use of such rice shoots. If new shoots growing from the stubble of a harvested first crop or a crop of early-maturing rice are rejuvenated by an application of fertilizer, a goodly quantity of regenerated rice might be reaped from a quarter-acre.

Surely nothing could be better than growing a biannual crop or two crops successively rather than having to repeatedly transplant. Why should we cling to the narrow view of rice as an annual crop that is sown in the spring and harvested in the fall? Although I have been intrigued by the possibility of harvesting rice twice after one seeding or even overwintering it and growing it as a perennial, I have not yet succeeded in finding a practical way to do this. I believe, however, that the idea definitely warrants investigation in warmer parts of Japan and in certain other countries.

The conclusions of natural farming were evident from the start, but it was achieving these in practice that took so long. I had to spend many years observing in order to understand the conditions under which rice seed will overwinter. And even if I understood why it would not overwinter in a particular instance and was able to eliminate the reasons, I preferred not to use scientific means or pesticides. I pondered too the meaning and worth of cultivating perennial rice.

Natural farming does not treat the planting of seed separately, but relates it to all other aspects of rice production. In contrast, scientific farming divides rice cultivation into narrow specialties; experts on germination attend to problems of seed germination, specialists in tillage address tilling problems, and likewise with direct seeding, transplanting, and other areas.

Natural farming treats everything as part of a whole. The problems may differ, but solving them independently is totally meaningless. In rice cultivation, preparing the field, sowing the seed, tilling, covering the seed with soil, fertilizing, weeding, and disease and pest control are all organically interrelated. No problem in any one area is truly solved unless a common solution is found for all areas.

One thing is all things. To resolve one matter, one must resolve all matters. Changing one thing changes all things. Once I made the decision to sow rice in the fall, I found that I could also stop transplanting, and plowing, and applying chemical fertilizers, and preparing compost, and spraying pesticides.

Biennial cultivation proved to be both a step forward and a step back because I had to decide first whether to transplant or to seed the fields directly.

Natural Direct Seeding: I began studying direct seeding when I realized that all plants in nature seed directly. It occurred to me that, the transplantion of rice seedlings being a human invention, natural rice cultivation must involve direct seeding. So I tried sowing rice seed in the autumn. But my seed did not survive the winter and the attempt was a total failure. The reason was perfectly clear. Modern rice and other cultivated grains have been genetically improved for centuries; they are no longer natural and can never return to nature. In fact, sowing today's improved seed by a method that approximates nature is unnatural in itself. These plants require some form of protection and human care.

Yet, making use of an unnatural method of cultivation just because a cultivar is unnatural only moves the rice even further away from nature and evokes stronger natural repercussions. The grain was no longer natural, yet there had to be a more natural way to grow it. In addition to which, simply giving up all attempts because "overwintering rice seed is difficult" and "barley cannot be carried through the summer" would have ended the matter then and there without the least hope of getting an insight into the deepest designs of nature. So I set my sights on learning why rice does not overwinter.

In 1945, before I had gotten very far on this, I ran a different experiment in which I direct-seeded onto a plowed and flooded paddy field in the spring. I followed the same procedure as for preparation of a rice nursery bed, first plowing the field, then flooding and tilling it. After this was done, I seeded directly.

The experiment consisted of drilling, seeding in straight rows, and broadcasting. The main object was to examine the effects of different sowing techniques and the sowing rate density. I planted approximately 20, 30, 60, 100, 230, and 1000 seeds individually per square yard. The results were pretty much as I had expected and yet surprising. Aside from the extremely dense planting, the number of heads per square yard was about 400–500 in all cases, and the number of grains per head from 60 to 120. Yields were therefore about the same.

Several problems did arise. For example, where the soil was rich in organic matter and bad water collected, the seed sunk into the ground and germination was poor. I also noticed that deep flooding of the field resulted in plants that tended to lodge easily. But, all in all, rice generally grew well when direct-seeded on plowed and irrigated paddy.

I spent so much time weeding that I doubt this method had much practical value at the time. But with the good herbicides around today, direct seeding on an unplowed, poorly drained, or moderately drained field is definitely possible.

Early Attempts at Direct-Seeding, No-Tillage Rice/Barley Succession

I tried many different ways of direct seeding, but since the method I used initially to plant the preceding barley crop was to drill seeds on high ridges, I picked up the idea of drilling the rice seed in the furrows between the ridges from a "lazy man's" method of sowing attempted by some farmers long ago. This led to a later technique I used of direct-seeding rice between rows of barley.

I direct-seeded rice between barley for several years, but I had so much trouble with rice germination and weed control that I finally gave this method up as impractical. During this period, however, I was experimenting with many other methods, which gave me some fresh ideas. Here are a few of the things I tried.

First trial: Direct seeding of rice between barley
1) Germination of the rice seed was poor. There was no way to fight off mole crickets, sparrows, and mice. I tried using pesticides, but was unable to achieve full germination.
2) After harvesting the barley, I tried intertilling the soil on the ridges with a hoe, and also leveling the field by transferring ridge soil into the furrows between the ridges, but this was arduous work.
3) Even when I irrigated the fields, water retention was poor and weeds grew on high ridge areas exposed above the surface of the water. I had a great deal of trouble dealing with weeds along the water's edge and in the water, and with the complicated pattern of weed emergence. Use of herbicides was more difficult than for transplanted rice, which further complicated weed control.
4) Finally, after having pondered over the best way to weed, I thought of controlling weeds with weeds, and tried sowing the clover and Chinese milk vetch that I was experimenting with in my orchard over the ridges of maturing barley one month before the barley harvest so as to get a rich growth of these herbs among the barley. This method was not immediately successful, but it gave me another important clue that was to lead later on to my method of rice and barley cropping in a ground cover of clover.
5) I tried sowing vegetable seeds, such as mustards, beans, and squash, and although none of these grew well enough to be of much use for home consumption, this taught me something about the relationships between specific crops in a rotation.
6) I then tried the opposite: seeding and growing rice in fields of tomato, eggplant, and cucumber. Rice yields were better here than my attempts at raising vegetables in a rice paddy and growing rice after harvesting the vegetables, although I did have some problems with field work.

Second trial: Direct-seeding rice/barley succession
I mentioned earlier that because my research on the direct seeding of rice on

drained fields was tied in with the direct seeding of barley, as my method of barley cropping progressed from high-ridge to low-ridge to level-field cultivation, my method of direct seeding rice followed suit, moving toward level-field, direct-seeding cultivation. From seeding in single rows at wide, 18-inch intervals, I went to planting in narrowly spaced rows 6 to 8 inches apart, then to planting seeds individually at intervals of 6 by 8 inches, and finally I direct-seeded naked barley over the entire surface of the field without plowing or tilling.

This was the start of the no-tillage direct-seeding of naked barley. Because my method resulted in the high-yield cultivation of barley and the dense individual planting of seed, I found it increasingly difficult to sow rice seed among the barley. One reason was the lack of a planter at the time that could seed effectively between barley plants.

I had learned therefore that naked barley can be grown quite well by sowing seeds individually on a level, unplowed field. Having also found that rice sown at the same seeding interval among the barley stubble grows very well, it dawned on me that, since I was using exactly the same method for growing both rice and barley, and was growing these two crops in succession one after the other, both crops could be grown as a single cropping system. I chose to call this system "direct-seeding, no-tillage rice/barley succession."

However, this system was not the result of a sudden flash of inspiration. It was the outcome of many twists and turns. When I learned the inconvenience of direct-seeding rice between barley stubble, I decided to run tests to determine whether to direct-seed rice after harvesting the barley or to broadcast the rice seed over the heads of barley ten to twenty days before cutting the barley.

Scattering rice seed over the standing heads of barley is truly an extensive method of cultivation, but seed losses due to sparrows and mole crickets were lighter than I had expected and percent germination quite good. Although I thought this to be an interesting method, I practiced it only in one corner of my field and did not pursue it any further at the time, preferring instead to concentrate on the direct-seeding of rice following the barley harvest.

I did make an attempt to plant rice seed directly onto the harvested barley field without plowing, but this did not work out well with the planter and the rice seed merely fell to the ground resulting in a shallow planting depth. I remember feeling then that sowing the rice seed over the standing barley would have been preferable, but for various reasons having to do with the method of cultivation and ease of lodging, I decided to try direct seeding on a shallow-tilled field instead. Also, because I continued to believe at the time that the most important condition for high barley and rice yields was deep plowing, I felt that tilling was a necessary precondition for the direct seeding of rice.

But direct-seeding with shallow plowing turned out to be more difficult than I thought, for it required harrowing and leveling just as in the preparation of a seed bed for rice. And the risks are very great, especially in only partially drained fields and during years of abundant rainfall. If rain falls on the plowed field before seeding, the field turns to mud, making direct seeding impossible. After repeated failures over a number of years, I decided to go with the principle of direct seeding without tilling of any sort.

176

Third trial: Direct-seeding, no-tillage rice/barley succession

Today I use the term "direct-seeding, no-tillage rice/barley succession" without thinking twice about it, but until I was fully convinced that the field does not have to be plowed or worked, it took incredible resolve for me to say "no-tillage" and propose this method of cultivation to others.

This was at a time when, despite scattered attempts to "half-plow" wheat or adopt simplified methods of preparing the rice field for planting, the conventional wisdom held deep plowing to be necessary and indispensable for producing high yields of both rice and barley. To abstain from plowing and tilling a field year after year was unthinkable.

I have grown rice and barley without any plowing for well over ten years now. My observations during that period, coupled with other insights, have gradually deepened my conviction that the paddy field does not need to be plowed. But this conviction is based largely on observation, as I have not conducted studies and collected data on the soil. Yet, as one soil scientist who examined my field put it, "A study can look at the changes that arise with no-tillage farming, but it can't be used to judge the merits of no-tillage farming based on conventional ideas."

The ultimate goal is the harvest. The answer to this question of merit depends on whether rice yields decline or increase when no-tillage farming is continued. This is what I wanted to find out. At first, I too expected that yields would drop off after several years of continuous no-tillage farming. But perhaps because I returned all rice and barley straw and hulls to the land, during the entire period that I have used this method, I have never seen any sign of a decline in yields due to reduced soil fertility. This experience sealed my conviction that no-tillage farming is sound in practice and led me to adopt this as a basic principle of my farming method.

In 1962, I reported these experiences of mine in an article entitled "The Truth about Direct-Seeding Rice and Barley Cultivation," published in a leading farming and gardening journal in Japan. This was regarded as a highly singular and unconventional contribution, but apparently acted as a strong stimulus on those interested in the direct seeding of rice. One high-ranking official in the Ministry of Agriculture and Forestry at the time was delighted and encouraging, calling it "research in a class by itself . . . a guiding light for Japanese rice cultivation ten years hence."

Natural Rice and Barley/Wheat Cropping

I adopted the standpoint of natural farming early on, and discontinuing the transplantation of rice, sought my own method of rice and barley direct seeding. In the process, I gradually approached a unified technique of direct-seeding naked barley and rice without tilling that brought me a step closer to my goal. This can be thought of as the antecedent of the direct-seeded upland rice cropping methods practiced widely today. At the time, nobody would have thought that rice and naked barley could be grown on a level field continuously left unplowed.

Later, as a result of determined efforts to reject the use of pesticides and fertilizers, I began a method of cultivation in keeping with my goal of natural farming: a very simple form of continuous, no-tillage rice/barley cropping involving direct seeding and straw mulching. I adopted this as the basic pattern for natural farming.

This method was studied at a large number of agricultural testing stations throughout Japan. In almost every instance, researchers found there to be no basic problem with the no-tillage, succession cropping of rice and barley using straw mulch. But weed control remained a problem, so I worked on this and after a great deal of effort and repeated experimentation, modified my basic method by adding a ground cover of green manure, the mixed seeding of rice and barley, and biennial cultivation.

I called this the basic pattern of natural rice and barley farming because I was certain that this technique enabled the farmer for the first time to farm without using any pesticides or chemical fertilizers. And I referred to it also as the "clover revolution" in rice and barley cropping to voice my opposition to modern scientific farming with its use of chemicals and large machinery.

Direct-Seeding, No-Tillage Barley/Rice Succession with Green Manure Cover

This is a method for the companion cropping of leguminous green manure plants with rice and barley or wheat, all members of the grass family.

Cultivation Method: In early or mid-October, I sow clover seeds over the standing heads of rice, then about two weeks before harvesting the rice, I sow barley seed. I harvest the rice while treading over the young barley seedlings, and either dry the cut grain on the ground or on racks. After threshing and cleaning the dried grain, I immediately scatter the rice straw uncut over the entire field and apply chicken manure or decomposed organic matter. If I wish to overwinter my rice, I enclose the rice seed in clay pellets and scatter these over the field in mid-November or later. This completes the sowing of rice and barley for the coming year. In the spring, a thick layer of clover grows at the foot of the maturing barley, and beneath the clover, rice seedlings begin to emerge.

When the barley is cut in late May, the rice seedlings are perhaps an inch or two high. The clover is cut together with the barley, but this does not interfere with the harvesting work. After leaving the barley on the ground to dry for three days, it is gathered into bundles, then threshed and cleaned. The barley straw is scattered uncut over the entire field, and over this, a layer of chicken manure is spread. The trampled rice seedlings emerge through this barley straw and the clover grows back also.

In early June, when the rich growth of clover appears about to choke out the young rice seedlings, I plaster the levees around the field with mud and hold water in the field for four to seven days to weaken the clover. After this, I surface-drain the field in order to grow as hardy plants as possible. During the first half of the rice growing season, irrigation is not strictly necessary, but depending on how the plants are growing, water may be passed briefly over the field once every week to ten days. I continue to irrigate intermittently during the heading stage, but make it a point not to hold water for more than five days at a stretch. A soil moisture level of eighty percent is adequate.

During the first half of its growing season, rice does well under conditions similar to those in upland rice cultivation, but in the second half of the season, irrigation should be increased with plant growth. After heading, the rice requires lots of water and without careful attention could become dehydrated. For yields of about one ton per quarter-acre, I do not make use of standing water, but careful water management is a must.

Farmwork: This method of rice cultivation is extremely simple, but because it is a highly advanced technique, quite unlike extensive farming, each operation must be performed with great precision. Here is a step-by-step description of the operations, starting at the time of rice harvest in the fall.

1. Digging drainage channels: The first thing one has to do when preparing a normal paddy field for the direct-seeded no-tillage cropping of rice and barley is to prepare drainage channels. Water is normally held in the paddy throughout the rice growing season, turning the soil to a soft mud. As harvest time approaches, the surface must be drained and dried to facilitate harvesting operations. Two or three weeks before the rice is cut, a water outlet is cut through the levee surrounding the field and the surface of the field drained. A row of rice about the perimeter is dug up with a cultivator, transferred inwards out of the way, and a drainage channel dug.

For good drainage, the channel must be dug deeply and carefully. To do this, make a furrow in the soil with the end of a long-handled sickle, dig up the rice plants along the furrow, then shape a channel about 8 inches deep and 8 inches wide by lifting the soil aside with a hoe.

After the rice has been harvested, dig similar drainage channels in the field at intervals of 12 to 15 feet. These provide sufficient drainage to enable good growth of green manure crops and barley even in a moist field. Once dug, these drainage channels can be used for many years in both rice and barley cultivation.

2. Harvesting, threshing, and cleaning the rice: Cut the rice while trampling over the clover and the young, two- to three-leaf barley shoots. Of course, the rice may be harvested mechanically, but where the size of the field permits, it is both sufficient and economical to harvest with a sickle and thresh with a pedal-powered drum.

3. Seeding clover, barley, and rice:

Seeding method: When seeded over the standing heads of rice, the clover and barley seed readily germinate because of the high soil moisture. Winter weeds have not yet appeared, so this is helpful for controlling weeds. The barley and rice seed may be drilled or sown individually in straight rows following the rice harvest, but broadcasting directly over the maturing heads of rice requires less work and is beneficial for germination, seedling growth, and weed control.

Seeding date and quantity per quarter-acre:

Clover	1 lb.	September-October and March-April
Barley	6.5–22 lbs.	end of October to mid-November
Rice	6.5–22 lbs.	mid-November to December

When aiming for high yields, it is a good idea to seed sparsely and evenly, but seed 22 pounds each of rice and barley initially.
Variety: For normal yields, use varieties suited to your area, but for high yields, use hardy, panicle weight type varieties with erect leaves.

Overwintering rice: The seed will have to be coated. Seeds coated with a synthetic resin solution containing fungicide and pesticide and sown in the autumn will survive the winter. To eliminate the use of pesticides, enclose the seeds in clay pellets and scatter the pellets over the field.

Preparing the clay pellets: The simplest method is to mix the seeds in at least a five- to ten-fold quantity of well-crushed clay or red earth, add water, and knead until hard by treading. Pass the kneaded mixture through a half-inch screen and dry for a half-day, then shape the clay mixture into half-inch pellets by rolling with the hands or in a mixer. There may be several (4–5) seeds in each pellet, but with experience this can be brought closer to the ideal of one seed per pellet.

Table 4.4   Growing seasons for direct-seeded rice and barley/wheat cultivation.

Cultivation Method	Previous Crop	Nov.	Dec.	Jan.	Feb.	Mar.	Apr.	May	Jun.	Jul.	Aug.	Sep.	Oct.	Nov.	Rice Crop	
(1) Direct-seeding rice after barley/wheat harvest	Naked barley	○						× ○					×		Early	Late
	Wheat	○						× ○						×	Early	Late
(2) Direct-seeding rice among maturing barley/wheat	Naked barley Wheat	○						×					×		Early	Late
						○○										
(3) Simultaneous direct-seeding of rice and barley/wheat (autumn)	Naked barley (early)	○○					×						×		Early	(Late)
		○○														
(4) Winter/spring direct-seeding of rice	Autumn vegetables		○○					× × ×							Early	(Late)
				○○									×			
(5) Direct-seeding of rice and barley/wheat in ground cover of clover	Naked barley Clover	○					×	○					×			Late

○······Planting date   ×······Harvesting date

To prepare one-seed pellets, place the seed moistened with water in a bamboo basket or a mixer. Sprinkle the seed with clay powder while spraying water mist onto the mixture with an atomizer and moving the basket in a swirling motion. The seeds will become coated with clay and grow larger in size, giving small pellets a quarter- to a half-inch in size. When a large quantity of pellets is to be prepared, one alternative is to do this with a concrete mixer.

Topsoil-containing clay may also be used to form the pellets, but if the pellets crumble too early in spring, the seed will be devoured by rodents and other pests. For those who prefer a scientific method of convenience, the seeds may be coated with a synthetic resin such as styrofoam containing the necessary pesticides.

Single cropping: Even when rice is single-cropped rather than grown in alternation with barley, clover seed may be sown in the fall, and the following spring rice seed scattered over the clover and the field flooded to favor the rice. Another possibility is to sow Chinese milk vetch and barley early, then cut these early in spring (February or March) for livestock feed. The barley will recover enough to yield 11 to 13 bushels per quarter-acre later. When single-cropping rice on a dry field, bur clover or Chinese milk vetch may be used.

Shallow-tillage direct-seeding: Twenty-two pounds each of barley and rice seed may be sown together in the autumn and the field raked. An alternative is to lightly till the field with a plow to a depth of about two inches, then sow clover and barley seed and cover the seed with rice straw. Or, after shallow tilling, a planter may be used to plant seed individually or drill. Good results can be had in water-leak paddy fields by using this method first, then later switching to no-tillage cultivation. Success in natural farming depends on how well shallow, evenly sown seeds germinate.

4. Fertilization: Following the rice harvest, spread 650–900 pounds of chicken manure per quarter-acre either before or after returning the rice straw to the fields. An additional 200 pounds may be added in late February as a topdressing during the barley heading stage.

After the barley harvest, manure again for the rice. When high yields have been collected, spread 450–900 pounds of dried chicken manure before or after returning the barley straw to the field. Fresh manure should not be used here as this can harm the rice seedlings. A later application is generally not needed, but a small amount (200–450 pounds) of chicken manure may be added early during the heading stage, preferably before the 24th day of heading. This may of course be decomposed human or animal wastes, or even wood ashes.

However, from the standpoint of natural farming, it would be preferable and much easier to release ten ducklings per quarter-acre onto the field when the rice seedlings have become established. Not only do the ducks weed and pick off insects, they turn the soil. But they do have to be protected from stray dogs and hawks. Another good idea might be to release young carp. By making full, three-dimensional use of the field in this way, one can at the same time produce good protein foods.

5. Straw mulching: Natural rice farming began with straw. This promotes seed germination, holds back winter weeds, and enriches the soil. All of the straw and chaff obtained when harvesting and threshing the rice should be scattered uncut over the entire surface of the field.

Barley straw too should be returned to the field after the harvest, but this must be done as soon as possible following threshing because once dried barley straw is wet by rain, it becomes more than five times as heavy and very difficult to transport, in addition to which the potassium leaches out of the straw. Often too, attempting to do a careful job can be self-defeating, for with all the trouble it takes to get out the cutters and other motorized equipment, one is often tempted to just leave the straw lying about.

No matter how conscientious a farmer is in his work, each operation is part of a carefully ordered system. A sudden change in weather or even a small disruption in the work schedule can upset the timing of an operation enough to lead to a major failure. If the rice straw is scattered over the field immediately after threshing, the job will be done in just two or three hours. It does not really matter how quick or carelessly it is done.

Although it may appear to be crude and backward, spreading fresh straw on a rice field is really quite a bold and revolutionary step in rice farming. The agricultural technician has always regarded rice straw as nothing but a source of rice diseases and pests, so the common and accepted practice has been to apply the straw only when fully decomposed as prepared compost. That rice straw must be burned as a primary source of rice blast disease is virtually accepted doctrine in some circles, as illustrated by the burning of rice straw on an immense scale in Hokkaido under the urging of plant pathologists.

I deliberately called composting unnecessary and proposed that all the fresh rice straw be scattered over the field during barley cultivation and all the barley straw be spread over the field during rice cultivation. But this is only possible with strong, healthy grain. How very unfortunate it is then that, overlooking the importance of healthy rice and barley production, researchers have only just begun to encourage the use of fresh straw by chopping part of the straw with a cutter and plowing it under.

Straw produced on Japanese rice fields is of great importance as a source of organic fertilizer and for protecting the fields and enriching the soil. Yet today this practice of burning such invaluable material is spreading throughout Japan. At harvest time in the early summer, no one stops to wonder about the smoke hanging over the plain from the burning barley straw in the fields.

A number of years ago, a group of farming specialists and members of the agricultural administration, most of whom had no first-hand idea of how much hard work preparing compost is, did start a campaign urging farmers to enrich the soil by composting with straw. But today, with the large machinery available, all the harvesting gets done at once. After the grain has been taken, the problem for many seems to be how to get rid of all the straw; some just let it lay and others burn it. Are there no farmers, scientists, or agricultural administrators out there who see that whether or not we spread straw over our fields may decide the fate of our national lands?

It is from just such a small matter that shall emerge the future of Japanese agriculture.

6.   Harvesting and threshing barley: Once the barley has been seeded and the mulch of rice straw applied, there is nothing left to do until the barley is ready for harvesting. This means one person can handle whatever needs to be done on a quarter-acre until harvest time. Even including harvesting and threshing operations, five people are plenty for growing barley. The barley can be cut with a sickle even when broadcast over the entire field. A quarter-acre will yield over 22 bushels (1,300 pounds).

7.   Irrigation and drainage: The success of rice and barley cropping depends on germination and weed control, the first ten to twenty days being especially critical.

Water management, meaning irrigation and drainage, is the most important area of crop management in rice cultivation. Irrigation management throughout the rice growing season can be particularly perplexing for the novice farmer, and so merits special attention here.

Farmers making use of these methods of direct-seeding rice-barley cultivation in areas where most farmers transplant their rice will be seeding and irrigating at times different from other local farmers. This can lead to disputes, especially as the irrigation canals are communally controlled; one cannot simply draw large amounts of water from a long canal whenever one pleases. Also, if you irrigate when the neighboring fields are dry, water leakage into other fields can greatly inconvenience the farmer next door. If something like this happens, immediately plaster your levees with mud. With intermittent irrigation, fissures tend to develop in the levee, causing leakage.

Then too there is always the problem of moles. Most people might dismiss a mole tunnel as nothing much to worry about, but a mole running along the length of a freshly plastered levee can in one night dig a tunnel 40–50 feet long, ruining a good levee. By burrowing straight through a levee, a mole weakens it so that water even starts leaking out of mole cricket and earthworm holes; before you know it, a sizable hole has formed. Finding holes in levees may appear to be easy, but unless the grass along the top and sides of the levee is always neatly cropped (it should be cut at least three times a year), there is no way of knowing where the entrance or exit is. More often than not, one notices a hole for the first time only after it has widened up considerably.

A hole may appear small from the outside, but inside it widens into larger pockets that just cannot be stopped up with a handful or two of mud. If dirt has flowed out of a hole for an entire night, you will have to carry in maybe 50 to 100 pounds of earth to repair it. Use stiff earth to plug up the hole; if it is plugged with soft earth, this might work free overnight. Avoid makeshift repairs as these only lead to eventual crumbling of the ridge, which will spell real trouble.

Do not leave grass cuttings and bundles of straw on a levee as these draw earthworms which moles come to feed on. If moles are present, they can be gotten rid of using a number of devices. For example, these can be caught merely by placing a simple bamboo tube capped at both ends with valves at a hard point in the mole tunnel. There is a trick to catching moles, but once you have gotten the knack of it and are finally able to keep your entire field filled with water by plugging all the holes, then you too will be a full-fledged rice farmer.

After having experienced the tribulations of water management, you will be better prepared to fully appreciate the hardships and rewards of natural farming.

Lately, highland paddy rice farmers have been constructing their levees of concrete or covering the footpaths with vinyl sheeting. This appears to be an easy way of holding water, but the earth at the base of the concrete or below the sheeting are ideal places for moles to live. Give them two or three years and repairs on these might be a lot more difficult than on normal earthen levees. In the long run, such methods do not make things easier for the farmer.

All one needs to do, then, is to rebuild the levees each year. To build a levee that does not leak, first carefully cut the grass on the levee with a sickle, then break down the levee with an open-ended hoe. Next, dig up the soil at the bottom of the levee and, drawing some water alongside, break up and knead the earth with a three-pronged cultivator. Now build up the levee and, after letting this stand for awhile, plaster the top and sides with earth.

All the traditional farming tools used from ancient times in Japan come into play during the building of an earthen levee. Observing the processes by which these simple yet refined implements efficiently modify the arrangement of soil particles in the paddy field, I get a keen sense of just how perfectly designed and efficient they are. Even in soil engineering terms, these tools and their use represent a highly refined technology.

Such a technology is clearly superior to poured concrete and vinyl sheeting. Erecting a well-built levee in a paddy field is akin to making a work of art. Modern man sees the mud-coated farmer plastering his levees and transplanting his rice as a throwback to a crude, prescientific age. The mission of natural farming is to peel away this narrow vision and show such labor in its true light as artistic and religious work.

8. *Disease and pest "control"*:   After twenty to thirty years of farming without pesticides, I have come to believe that, while people need doctors because they are careless about their health, crops do not indulge in self-deception and, provided the farmer is sincere in his efforts to grow healthy crops, will never have any need for pesticides.

To the scientific skeptics, however, the matter is not so easily settled. Yet my years of experience have shown me the answers to their doubts and pointed questions. Questions such as: Wasn't that just a chance success? Why, you had no large outbreak of disease or pest damage, did you? Aren't you just benefiting from the effects of pesticides sprayed by your neighbors? Aren't you just evading the problem? So where do the pests go, then?

There have been massive local outbreaks of leafhoppers on two or three occasions over the past thirty years, but as the records of the Kochi Prefecture Agricultural Testing Station bear out, no ill came of a lack of control measures. No doubt, if such surveys were conducted on a regular basis year in and year out, people would be more fully convinced. But of even greater importance, certainly, is the knowledge of just how complex and filled with drama is the world of small creatures that inhabit a rice field.

I have already described just how profound are the effects of pesticides on a living field. My field is populated with large Asiatic locusts and tree frogs; only over this field will you find hovering clouds of dragonflies, and see flocks of ground sparrows and even swallows flying about.

Before we debate the need to spray pesticides, we should understand the dangers posed by man's tampering with the world of living things. Most damage caused by plant diseases and pests can be resolved by ecological measures.

High-Yield Cultivation of Rice and Barley

Many people assume that yields from natural farming are inferior to those of scientific farming, but in fact the very reverse is true.

Analytic and scientific reasoning leads us to believe that the way to increase yields is to break up rice production into a number of constituent elements, conduct research on how to make improvements in each, then reassemble the elements once they have been improved. But this is just like carrying a single lantern to guide one's way through a pitch-dark night. Unlike one who makes his way without a lantern toward the single, faraway light of an ideal, this is blind, directionless progress. The scientific research from which technology unfolds lacks a unity of purpose; its aims are disparate. This is why techniques developed through research on rice that yields 15 bushels per quarter-acre cannot be applied to rice that gives 30 or 40 bushels. The quickest and surest way to break through the 20-bushel barrier is to take a look at 30- or 40-bushel rice, and setting a clear goal, concentrate all one's technical resources in that direction.

Once the decision has been made to go with rice plants having a given panicle-to-stalk length ratio such as 8:1, 6:1, or 3:1, say, this clarifies the goal for farmers producing the rice, enabling the shortest possible path to be taken towards achieving high yields.

The Ideal Form of a Rice Plant: Aware of the inherent problems with the process of breaking down and analyzing a rice plant in the laboratory and reaching conclusions from these results, I chose to abandon existing notions and look instead at the rice plant from afar.

My method of growing rice may appear reckless and absurd, but all along I have sought the true form of rice. I have searched for the form of natural rice and asked what healthy rice is. Later, holding on to that image, I have tried to determine the limits of the high yields that man strives after.

When I grew rice, barley, and clover together, I found that rice ripening over a thick cover of clover is short-stalked, robust right down to the bottom leaf, and bears fine golden heads of grain. After observing this, I tried seeding the rice in the fall and winter, and learned that even rice grown under terrible conditions on arid, depleted soil gives surprisingly high yields.

This experience convinced me of the possibilities of growing high-yielding rice on continuously untilled fields, so I began experimenting to learn the type of field and manner in which rice having an ideal form will grow. Eventually I found what I thought to be the ideal form of high-yielding rice. Tables 4.5 through 4.7 give the dimensions of ideal rice. Each value indicated is the average for three plants.

Table 4.5   Dimensions of ideal rice plants.

(Units: inches)

	Cultivar:	A	B	C
Head length		6.9	6.5	5.9
Internode length	1st	9.4	9.6	9.1
	2nd	5.3	6.1	6.3
	3rd	4.3	3.9	5.1
	4th	1.2	2.4	2.8
	5th	0	0	1.2
Stalk length		20.2	22.0	24.5
Leaf blade length	1st	9.1	8.7	8.3
	2nd	11.4	12.2	11.4
	3rd	9.8	15.7	14.2
	4th	7.5	16.5	15.0
	5th	—	—	11.8
Total		37.8	53.1	60.7
Leaf sheath length	1st	9.4	9.1	8.7
	2nd	7.1	7.1	6.7
	3rd	6.5	7.1	6.7
	4th	5.5	7.5	7.1
	5th	—	—	6.3
Total		28.5	30.8	35.5

Table 4.6   Stalk length and first internode length.

(Units: inches)

	Cultivar:	A	B	C
Stalk length (S)		20.3	22.0	24.4
First internode length (F)		9.4	9.6	9.1
Ratio (F/S × 100)		46	44	37

Table 4.7   Length of leaf blade + leaf sheath.

(Units: inches)

	Cultivar:	A	B	C
First leaf		18.5	17.7	16.9
Second leaf		18.5	19.3	18.1
Third leaf		16.1	22.8	20.9
Fourth leaf		13.0	24.0	22.0
Fifth leaf		—	—	16.1

Analysis of the Ideal Form: What follows is a description of the major characteristics of rice plants with an ideal form.

1.   Short-stalked dwarf rice of robust appearance; leaves are short, wide, and erect. While Iyo-Riki rice is erect and short-stalked to begin with, this variety has an extremely short stalk, the stalk height being just 21 inches. Seen growing in the paddy field, its small size makes it appear inferior to rice plants in surrounding fields, although it does have about 15 to 22 tillers per plant. At maturity, the stalks are heavy with bright golden heads of grain.

2.   The weight of the unhulled grain is 150 to 167 percent that of the straw. In ordinary rice, this is less than 70 percent, and generally 40 to 50 percent. When a dried stalk of rice is balanced on a fingertip, the point of equilibrium is close to the neck of the panicle. In ordinary rice, this is located near the center of the stalk.

3.   The length of the first internode at the top of the plant is more than fifty percent of the stalk length, and when the plant is bent downward at the first node, the panicle extends below the base of the stalk. The longer the length of this first internode and the larger the ratio of this length to the overall stalk length, the better.

4.   An important characteristic is that the leaf blade on the second leaf down is longer than that of any other leaf. Thereafter, the leaf blade becomes shorter as one moves down the stalk.

5.   The leaf sheaths are relatively long, the longest sheath being that on the first leaf. The sheaths become progressively shorter on moving down the plant. The total leaf length, representing the sum of the leaf blade length and sheath length, is longest for the first and second leaves, and decreases downward. In rice that is not high-yielding, the lower leaves are longer, the longest being the fourth leaf.

6.   Only the top four nodes grow, and the fourth is at ground level or lower. When the rice is cut, the straw includes no more than the two or three nodes. Normal rice has five or six nodes, so the difference is startling. When the rice is harvested, four or five leaves remain alive, but seeing as the top three fully formed leaves alone are enough to yield more than 100 full grains per head, the surface area required for starch synthesis is less than would otherwise be expected. I would put the amount of leaf surface needed to produce one grain of rice at perhaps 0.1 square inch, no more.

7.   A good plant shape naturally results in good filling of the grain. Weight per thousand grains of unpolished rice is 23 grams for small-grained rice, and 24.5–25 grams for normal-grained rice.

8.   Even at a density of 500 stalks to the square yard, hardy, upright dwarf rice will show no decline in the number of grains per head or percent of ripened grains.

Ideal Shape of Rice:

1.   Both the plant height and length of the leafblades are much smaller than in ordinary varieties. This is no accident. I had for some time thought large plants unnecessary in rice production, and so endeavored to suppress rather than promote vegetative growth of the plant. I did not irrigate during the first half of the growing season and by applying fresh straw to the field checked plant response to a basal application of fertilizer. As it turned out, I was correct. I have come to believe that internodal growth between the fifth and sixth nodes should be suppressed. In fact, I even believe that rice can do fine with just three aboveground nodes.

2.   In ideally shaped rice, the internode lengths each decrease by half from the top to the bottom of the plant. Not only does this indicate steady, orderly growth of the rice, it also means that internodal growth occurs only starting at the young panicle formation stage.

3.   The long second leaf and the decreasing leaf length as one moves down the stalk is the exact reverse of what is generally thought to be the correct shape of rice, but I believe that this inverted triangular shape gives a rice plant that does well in the fall.

Fig. 4.6   Ideal shape of a rice plant.

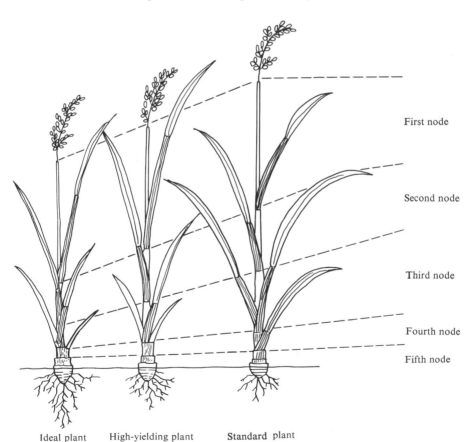

First node

Second node

Third node

Fourth node

Fifth node

Ideal plant   High-yielding plant   Standard plant

When all the leaves are erect, large top leaves give a better yield, but if the leaves are unhealthy and droop, highest yields are obtained with small, erect top leaves that do not shield the lower leaves from the sun. Thus, if plants with large upper leaves are grown but these leaves droop and yields decline as a result, this is because the rice plant is unhealthy and the lower leaves are too large.

4.   The leaf sheaths are longer than the leaf blades and enclose the stem of the plant. The long leaf sheath and blade on the flag leaf ensure the best possible nutritional state during the young panicle formation stage.

5.   After the seedling stage, the ideal rice plant remains small and yellow during the vegetative stage, but the leaves gradually turn greener during the reproductive stage. As measurements of the internode lengths show, changes in the nutritional state are steady and entirely unremarkable; fertilizer response increases with growth of the plant but never inordinately so.

Ideally then, the heads of rice are large and the plant short, having just three or four nodes above the ground. The leaves get longer in ascending order toward the top and the internode length between the fourth and fifth nodes at the bottom is very short. Instead of a feminine form with a high head-to-body ratio of six or even eight to one, this plant has a more sturdy, masculine, short-stalked, panicle weight type shape.

Of course, depending on the variety of rice, an ideal plant may have a long stalk and be of the panicle number type. Rather than deciding that some characteristic is undesirable, one should avoid producing weak, overgrown heads and strive always to practice methods of cultivation that suppress and condense. Concentrated rice carries a tremendous store of energy that provides high yields because it maintains an orderly shape receptive to sunlight, matures well, and is resistant to disease and pest attack—even in a very dense stand.

The next problem is how to go about growing an entire field of such rice.

Blueprint for the Natural Cultivation of Ideal Rice: Although raising one high-yielding rice plant with good photosynthetic efficiency is easy, it was no simple matter to grow stands of such rice.

A healthy individual rice plant growing in nature originally had plenty of space to grow. The sparse seeding of individual seeds allows the rice to assume the natural form that suits it best and to make full use of its powers.

Rice grown in its natural form puts out leaves in a regular, phyllotactic order. The leaves open up and spread in alternation, breaking crosswinds and ensuring the penetration of sunlight throughout the life of the plant, each leaf maintaining a good light-receiving form.

Knowing this, I anticipated from the start that healthy rice farming would require that I sow individual seeds sparsely. But because I was initially plagued with problems of poor germination and weed control when I began direct-seeded no-tillage cultivation, to ensure a stable crop I had no choice but to plant and seed densely.

However, dense planting and seeding tended to result in thick growth. The poor environment of individual plants made attempts to suppress growth ineffective, and the situation was doubly aggravated in wet years, when the rice would shoot up into tall, weak plants that often lodged, ruining the crop. To secure stable harvests of at least 22 bushels per quarter-acre, I had no choice but to resume sparse seeding. Fortunately, thanks to gradual improvements in the weed control problem and soil fertility, conditions fell into place that made it possible for me to seed sparsely. I tried broadcasting—a form of individual seeding, and also seeding at uniform intervals of from 6 to 12 inches. Tables 4.9 and 4.10 give my results.

Although I did run into a number of crop management problems, I found that sparse seeding gives healthy, natural rice plants that grow well and provide the high yields that I had expected. In this way, I was able to obtain yields of over one ton per quarter-acre with naturally grown rice. I should add that there is nothing absolute or sacred about the seeding rate and interval. These must be adjusted in accordance with other growing conditions.

The Meaning and Limits of High Yields: In natural farming, high yields rely on the absorption and storage of as much of nature's energy as possible by the crop. For this, the crop must make the fullest possible use of its inherent powers. The proper role of the natural farmer is not to utilize the animals and plants of nature so much as to help envigorate the ecosystem. Because crops absorb energy from the earth and receive light and heat from the sun, and because they use these to synthesize energy which they store internally, there are limits to the help man can provide. All he can do really is keep watch over the earth.

Rather than plowing the fields and growing crops, man would be better occupied in protecting the vitality of all the organisms inhabiting the earth and in guarding the natural order. Yet, it is always man who destroys the ecosystem and disrupts the natural cycles and flow of life. Call him the steward and keeper of the earth if you will, but his most important mission is not to protect the earth so much as to keep a close control over those who would ravage and waste it.

The guardian of a watermelon patch does not watch the watermelons, he looks out for watermelon thieves. Nature protects itself and sees to the boundless growth of the organisms that inhabit it. Man is one of these; he is neither in control nor a mere onlooker. He must hold a vision that is in unity with nature. This is why, in natural farming, the farmer must strictly guard his proper place in nature and never sacrifice something else to human desire.

Scientific farming consists of producing specific crops selected from the natural world to suit our human cravings. This interferes with the well-being of fellow organisms, setting the stage for later reprisal.

The scientist planning to cultivate high-yielding rice on a field sees the weeds growing at his feet only as pests that will rob sunlight and nutrients from the rice plants. He believes, understandably, that he will be able to achieve the highest possible yields by totally eradicating such "intruders" and ensuring that the rice plants monopolize the sun's incident rays. But removing weeds with herbicides is all it takes to upset the delicate balance of nature. The herbicides destroy the ecosystem of the insects and microorganisms dependent on the weeds, abruptly changing the current of life in the soil biocommunity. An imbalance in this living soil inevitably throws the organisms there off balance as well. Unbalanced rice is diseased rice, and therefore highly susceptible to concentrated attack by disease and insect pests.

Those who believe that the monopoly by rice, in the absence of weeds, of the sun's rays will provide the highest possible yields are sadly mistaken. Unable to absorb the full blessings of the sun, diseased rice wastes it instead. With its limited perception, scientific farming cannot make the same full use of solar energy as natural farming, which views nature wholistically.

Before pulling the weeds growing at the base of the rice plants, natural farming asks why they are there. Are these grasses the by-product of human action or did they arise spontaneously and naturally? If the latter, then they are without doubt of value and are left to grow. The natural farmer takes care to allow natural plants that protect the natural soil to carry out their mission.

Green manure thriving at the foot of the rice plants and, later, algae growing on the flooded field are thought to detract from yields because they directly and indirectly shield the sun, reducing the amount of light received by the rice plants.

But we reach a different conclusion if we see this as a nearly natural state. The total energy absorbed by the rice, green manure, algae, and earth is greater than the energy stored from the sun's rays by the rice plants. The true value of energy cannot be determined merely by counting the number of calories. The quality of the energy produced within the plant by conversion from absorbed energy must also be taken into account. There is a world of difference between whether we look only at the amount of energy received by the rice plant or take a three-dimensional view of its quantitative and qualitative utilization of energy from the sun's rays.

Table 4.8 Breakdown of harvest yields.

Cultivar:	A	B	C
Plants per square yard	20	20	20
Heads per plant	18	20	20
Ripened grains per head	115	70	53
Unripened grains per head	10	18	21
Range in total grains per head	90–150	62–128	56–116
Ripened grains per plant	2,070	1,400	1,060
Weight of unhulled rice per plant (grams)	55.9	38.5	28.6
Weight of unpolished rice per plant (grams)	47.6	32.2	24.4
Weight of straw per plant (grams)	33	46	45.6
Weight ratio of unhulled rice to straw (%)	167	83	62
Weight of unhulled rice per thousand grains (grams)	27	27.5	27
Weight of unpolished rice per thousand grains (grams)	23	23	23
Yield per quarter-acre (kg)	1,165	787	597
Yield per quarter-acre (lbs.)	2,568	1,735	1,316

Energy from the sun is absorbed by the green manure plants. When the field is flooded, these wither and die, passing on their nitrogen to algae, which in turn become a source of phosphate. Using this phosphate as a nutrient source, microbes in the soil fluorish and die, leaving nutrients that are absorbed by the roots of the rice plant. If man were able to comprehend all of these cycles of energy and elements at once, this would become a science greater than any other. How foolish to focus only on solar energy apart from the rest of nature and think that merely by examining the amount of starch synthesis in the leaves of rice plants, one can gauge utilization of the sun's energy.

People must begin by understanding the futility of knowing bits and pieces of nature, by realizing that a general understanding of the whole cannot be acquired through value judgments on isolated events and objects. They must see that the moment the scientist endeavors to attain high yields by using the energy of the wind or sun, he loses a wholistic view of wind power and sunlight, and energy efficiency declines. It is a mistake to think of the wind and light as matter.

I too raise rice and analyze its growth, but I never seek to attain high yields through human knowledge. No, I analyze the situation we have today, where man has upset the natural order of things and must work twice as hard to prevent harvest losses, and I try to encourage people to see the error of their ways.

True high yields come about only through the spirited activity of nature, never apart from nature. Attempts to increase production in an unnatural environment invariably result in a deformed and inferior crop. Yields and quality only appear to be high. This is because man can add or contribute nothing to nature.

Since the amount of solar energy that can be received by a field of rice is finite, there is a limit to the yields attainable through natural farming. Many believe that because man has the ability to conceive and develop alternative sources of energy, there are no absolute upper limits to scientific development and increases in harvest. But nothing could be further from the truth. The power of the sun is vast and unlimited when seen from the standpoint of Mu, but when made the object of man's wants and cravings, even the sun's power becomes small and finite. Science cannot produce yields that exceed those possible through nature. Effort rooted in human knowledge is without avail. The only course that remains is to relinquish deeds and plans.

The question of whether the method of cultivation I propose, a direct-seeding no-tillage rice/barley succession in a ground cover of green manure, is a true prototype of nature must be judged according to whether it is a methodless method that approaches closer to nature.

Rice being best suited to Japanese soil as a first crop, and barley or wheat as a second crop, a successive cropping of rice and barley or wheat that provides a large total caloric output makes good use of Japanese land by utilizing the full powers of nature.

The reason I concentrated on a method of biennial cultivation that begins by sowing rice seed in the autumn and devotes a full year to the growth of rice was because I thought that this would enable the rice to absorb the most natural energy throughout the year.

The cover of green manure makes three-dimensional use of space in the field,

Table 4.9 Blueprint for high-yield rice cultivation.

Category	Target yield* (kg/1/4-acre)	Seeding rate (kg/1/4-acre)	Germinated seeds** per m²	Spacing/seed*** (cm²)	Tillers per plant		Total heads per m²		Grains per head		Total grains per m²		Remarks
					Extra-heavy panicle type	Heavy panicle type	Extra-heavy panicle type	Heavy panicle type	Extra-heavy panicle type	Heavy panicle type	Extra-heavy panicle type	Heavy panicle type	
1	1,500	1	10	30	25	40	200	350	300	—	—	—	Extremely high yields
		1.4	15	27	20	30	250	400	270	—	68,000	—	
2	1,200	2	20	25	15	25	300	450	250	120	75,000	(5.4)	Intensively high yields
		3	30	17	12	20	350	500	200	110	70,000	(5.5)	
3	900	4	50	15	8	13	400	550	180	90	60,000	(5)	Stable high yields
		6	100	10	4	10	450	600	160	80	50,000	(4)	
4	750	8	250	6	2	3	500	650	150	70	50,000	(4)	Labor-saving cultivation
		12	500	4	1.5	1.5	600	700	140	60	40,000	(4)	
5	600	15	1,000	3	1	1	700	700	130	55	40,000	(4)	Extensive cultivation
		20	1,000	2	1	1	800	800	120	50	30,000	(3)	

*1 kg=2.2 lb.   **1 m²=1.2 yd²   ***1 cm²=0.155 in²

Table 4.10 Outline of rice cropping.

Category	Variety	Planting Time	Soil	Chicken Manure* (kg)	Water Management	Seeding Method
1	Extra-heavy panicle type	Autumn (Nov.–Dec.)	Rich soil	600 {(basal application—3, topdressing—1, during heading—2)	No standing water	Individual planting of seeds
2	Heavy panicle type	Winter (Dec.–Mar.)	Rich soil	500 (3, 0, 2)	No standing water	Planting of 1, 2, or 3 seeds at a time
3	Heavy panicle type or intermediate type	Spring (Apr.–May)	Normal soil	400 (2, 0, 2)	Intermittent irrigation	Planting of 1 to 6 seeds at a time
4–5	Same as above or panicle number type	Late-seeded (Jun.–Jul.)	Poor soil	300 (1, 0, 2)	Water-conserving cultivation	Broadcasting

*1 kg=2.2 lb.

Note (1) Extra-heavy panicle type----Happy Hill Nos. 2, 3; non-glutinous, glutinous
Heavy panicle type----Happy Hill No. 1; non-glutinous, glutinous
Intermediate type----Japanese and Korean heavy panicle types
Panicle number type----standard Japanese varieties
(2) This table also applies to planting during barley and wheat cropping.

while straw mulching and the breakdown of materials in the soil encourage revitalization of the natural ecosystem. These can be thought of as manifestations of an effort to approach the ultimate goal of a "do-nothing" nature. One look at the diagram in Fig. A at the beginning of this book, depicting the centripetal convergence of my research on rice cultivation, will make immediately clear what I have aimed at from the very beginning and where my efforts have brought me.

From a wholistic standpoint, the farming method I propose surely appears at least one step closer to nature. But to the scientist, this method is just one among many different ways of farming.

3. Fruit Trees

Establishing an Orchard

The same general methods used in reforestation can also be used to plant fruit trees and set up an orchard. Thus, rather than carting the trunks, branches, and leaves of felled trees off a contour-cleared orchard site, it makes more sense to arrange this material along contour lines and wait for it to decompose naturally.

The branches, leaves, and roots of the trees decompose after several years, becoming a source of organic fertilizer that supplies nutrients to the growing fruit trees. At the same time, a cover of organic matter helps to curb weed growth, prevents soil washout, stimulates the proliferation of microorganisms, and serves to enrich and otherwise improve the soil.

Fruit saplings should be planted at equal intervals along hill contours. Dig a fairly deep hole, fill it with coarse organic matter, and plant the sapling over this.

When starting up a natural farm, one should not clear and smooth the land with a bulldozer because this disturbs the humus-rich topsoil built up over a long period of time. Land developed with a bulldozer and left virtually bare for ten years is washed free of its topsoil, greatly shortening the economic life of the farm.

Because tree branches and leaves cut down when land is cleared interfere with farming operations, these are generally burned. But, like slash-and-burn agriculture, this sends the fertility of the land up in flames.

Tree roots that work their way down to the deepest soil strata contribute physically to the aggregation and structure of the soil. In addition, they also serve as a nutrient source and have a chelating action that solubilizes insoluble nutrients in the soil. If such valuable organic matter is dug up and disposed of when the land is cleared, this drastically changes natural conditions and so damages the soil that it is unable to recover, even if holes are later dug in the ground and the same amount of coarse organic matter returned.

194

In general, one foot of topsoil holds enough nutrients to sustain fruit trees for ten years without fertilization; similarly, three feet of rich soil can probably supply enough nutrients for about thirty years. If it were possible to use the rich, fertile soil of a natural forest in its natural form as a hot bed, cultivation without fertilizer might even be feasible.

People might expect tree growth and fruit harvests to suffer when fruit trees are planted without clearing the land at all, but in fact not only do these compare favorably, the economically productive lifetime of the land also tends to increase.

Natural Seedlings and Grafted Nursery Stock: After preparing the orchard soil, the next concern is planting. Obviously, from the standpoint of natural farming, we would expect trees grown from seed to be preferable to grafted nursery stock. The reasons usually given for planting grafted saplings are to make the plant early-bearing, to ensure consistent fruit size and quality, and to obtain early-ripening fruit. However, when a tree is grafted, the flow of sap is blocked at the graft juncture, resulting either in a dwarf tree that must be heavily fertilized, or in a tree with a short lifetime and poor resistance to temperature extremes.

When I tried the direct planting of mandarin orange seed, although I found that trees grown from seed are inferior and generally useless because they revert or degenerate, this gave me a clue as to the true form of the tree and its natural rate of growth. I will come back to this later.

While in principle a young tree grown from seed grows faster than grafted stock, I learned that when the initial grafted stock is one to two years old, natural seedlings do not grow as rapidly during the first two or three years and care is also difficult. However, when raised with great care, trees grown from seed develop the most quickly. Citrus rootstock takes more time and sends down shallower roots.

Citrus trees may generally be grown from nursery plants grafted with rootstock, which, although shallow rooted, are cold-hardy. Apple trees can be trained into dwarf trees by using dwarfing stock, but it may also be interesting in some cases to plant seed directly and grow the young saplings into majestic trees having a natural form. Such a tree bears fruit of vastly differing sizes and shape that are unfit for the market. Yet, on the other hand, there always exists the possibility that an unusual fruit will arise from the seed. Indeed, why not multiply the joys of life by creating a natural orchard full of variety and surprises?

Orchard Management: To establish a natural orchard, one should dig large holes here and there among the stumps of felled trees and plant unpruned saplings and fruit seed over the site, leaving these unattended just as one would leave alone a reforested stand of trees. Of course, suckers grow from the cut tree stumps and weeds and low brush flourishes. Orchard management at this stage consists primarily of coming in twice a year to cut the weeds and underbrush with a large sickle.

1. Correcting the tree form: Some pinching back is generally necessary on a young transplanted sapling to correct the arrangement of the branches. This is because, if dieback occurs at the tip or if too much of the root system has been cut, an unnaturally large number of suckers may emerge, causing the branches to become entangled. When the young tree lies in the shadow of a large tree, it tends to become leggy, in which case the lower branches will often die back. Left to itself, such a tree will acquire an unnatural form that will result in years of unending labor for the grower; to hasten the tree's approach to a more natural form, shoots and buds emerging from unnatural places must be nipped off as soon as possible.

Trees that show normal, steady growth right from the start assume a nearly natural form and can thereafter be left alone. Cutting the first one or two shoots is therefore very important. How well this is done can determine the shape of the tree over its entire lifetime and is a major factor in the success or failure of an orchard.

However, it is often hard to tell which shoots to leave and which to pinch off. The grower may decide, often prematurely, which branches are to be the primary scaffold branches and which the secondary scaffolds when the tree is still very young only to find later that these branches have tangled under other, unanticipated growth conditions. Early pruning can turn out to be unnecessary and even harmful when done unwisely.

It is all too easy to assume that a tree grown in a natural state will more easily acquire a natural form anyway, yet it is not through abandonment that a cultivated tree takes on a natural form, but only through the most careful attention and protection.

2. Weeds: I was especially interested in the growth and control of other trees and weeds in a natural orchard. Initially, four to five years after planting fruit trees, I found eulalia and other weeds growing thickly among the brush and assorted trees. Weeding was not easy and sometimes it was even hard to locate the fruit trees.

Although the growth of fruit trees among this other vegetation was irregular and yielded poor harvests in some cases, there was very little damage from disease and insects. I found it hard to believe that, with the odd assortment of trees in my orchard and some of the fruit trees even growing in the shadow of other trees, these were spared attack by diseases and pests.

Later, with continued cutting back of the underbrush, the non-fruit trees receded and weeds such as bracken, mugwort, and *kudzu* grew up in their place. I was able to control or suppress weed growth at this point by broadcasting clover seed over the entire orchard.

3. Terracing: Five to six years after planting, when the trees begin to bear fruit, it is a good idea to dig up the earth on the uphill side of the fruit trees with a hoe and construct terrace-like steps and a road on the orchard slope. Once these terraces have been built and the original weeds replaced, first with soft weeds such as chickweed, knotweed, and crabgrass, then with clover, the orchard begins to look like an orchard.

To create a natural orchard, one must observe the principle of the right crop for the right land. Hillside land and valley land must be treated as such. Avoid the monoculture of fruit trees. Plant deciduous fruit trees together with evergreen fruit trees and never forget to interplant green manure trees. These may include acacias which, as members of the pea family, produce nitrogenous fertilizer, myrtle—which produces nutrients such as phosphoric acid and potash, alder, and podocarpus.

You may also, with interesting results, interplant some large trees and shrubs, including climbing fruit vines such as grapevine, akebia, and Chinese gooseberry.

Leguminous green manure plants and other herbs that enrich the orchard soil may be planted as orchard undergrowth. Forage crops and semiwild vegetables can also be grown in abundance, and both poultry and livestock allowed to graze freely in the orchard.

A natural orchard in which full, three-dimensional use of space is made in this way is entirely different from conventional orchards that employ high-production techniques. For the individual wishing to live in communion with nature, this is truly a paradise on earth.

Building Up Orchard Earth without Fertilizers

The aim of soil management is to promote the conversion of weathered material from bedrock and stone into soil suitable for growing crops, and enrichment of this soil. The soil must be turned from dead, inorganic matter into an organic material that is alive.

Unfortunately, soil management as it is normally practiced today consists basically of clean cultivation that turns the soil into non-living mineral matter. Of course, there is a reason for this: repeated weeding, the application of chemical fertilizers, and careful management will increase yields and provide a good product.

The soil in many orchards has become depleted with constant plowing and weeding, so some farmers are hauling rice and barley straw from their paddy fields up into their hillside orchards and spreading it at the foot of fruit trees. This began more as a means of reducing weeding work than as a fundamental change in soil management. However, relying on straw from the field as the ground cover is hardly an ideal approach. All it does is keep the farmer busy hauling straw from the paddy up the hill and carrying weeds from the hillside down into the fields.

Soil management cut off from the field, garden, and hillside is meaningless; only a method that enriches all at the same time makes any sense.

Why I Use a Ground Cover: In order to make full use of the soil, soil management must be based on the use of a ground cover. This enables soil in the field, garden, and hillside orchard to become naturally enriched. It is far wiser to plant green manure trees and encourage the soil within the orchard to enrich naturally than to apply fertilizer.

When I set about to revive my father's orchard of old citrus trees following World War II, I began by studying soil conditioning, and especially ground cover cultivation, for the following reasons.

First of all, with all the topsoil washed away and only red clay remaining, passive efforts to reinvigorate the old trees by applying lots of fertilizer, root-grafting, and thinning blossoms would only have invited a further decline in the trees. Nor

Table 4.11   Herbs used as orchard cover crops.

Type of Herb	Growing Season	Uses
Grass Family		
Italian ryegrass orchardgrass	spring · summer	deciduous fruit tree undergrowth
timothy wild oats winter grains	summer/winter · spring	with fruit vines (control of summer weeds)
Pea Family		
common vetch hairy vetch* common vetch, Saatwicke*	winter · spring	evergreen trees, deciduous trees (control of spring weeds)
mung bean cowpea *kudzu*	spring · summer	large evergreen trees (control of summer weeds)
ladino clover* red/white clover alfalfa* crimson clover sweet clover sub clover	year-round	year-round weed control for all fruit trees
bur clover* Chinese milk vetch*	winter · spring spring	fruit trees and summer vegetables (control of spring weeds)
peanut* soybean* adzuki bean*	spring · summer	control of summer weeds (green manure)
lupine* broad bean* garden pea*	winter · spring	control of spring weeds (green manure)
Japan clover bush bean	spring	control of spring weeds
Mustard Family		
daikon turnip* Indian mustard* other mustards Chinese cabbage rapeseed* other vegetables	fall · winter	winter weed control for all fruit trees

*Important cover crops

would planting new saplings have worked any better since these would not have thrived in the poor soil.

The second reason was that, when looking at how my father had fared financially with the orchard, I found that the first thirteen years the orchard had been run at a loss, the next twenty years it had made money, and the follwing ten years were again run in the red. Even though the war had dealt the orchard a severe blow, still, I was amazed that what had at one time been regarded as one of the best local orchards had failed to make a net profit over more than forty years of operation.

Why? The answer is simple. While he celebrated his profit-making citrus crops, his sturdy trees, and his growing wealth, the orchard soil had become depleted.

I set out to raise fruit trees that grow as the soil enriches. This was one of the main reasons why I grew cover crops.

Ladino Clover, Alfalfa, and Acacia: What helps to rehabilitate depleted soil? I planted the seeds of thirty legumes, crucifers, and grasses throughout my orchard and from observations of these came to the general conclusion that I should grow a weed cover using ladino clover as the primary crop and such herbs as alfalfa, lupine, and bur clover as the secondary crops. To condition the deeper soil strata in the hard, depleted soil, I companion-planted fertilizer trees such as Morishima acacia, myrtle, and podocarp.

Features of Ladino Clover:

1) When used as a cover crop, this eliminates weeds. Annual weeds are displaced in one year, and biennials disappear in two years. After 2 to 3 years, almost all garden weeds have vanished, leaving a solid field of clover.
2) Improves soil down to a depth of 16 to 18 inches.
3) Seed does not have to be sown again for another 6 to 8 years.
4) Does not compete strongly with fruit trees for fertilizer or moisture.
5) Grows back easily after being cut, and remains healthy and hardy even when trampled upon.
6) Does not hinder farming operations.

The only disadvantages of ladino clover are that it is susceptible to summer-killing and sclerotium disease during hot, dry weather, and that growth is retarded in the shade and under trees.

Seeding Ladino Clover: The seed should be drilled the first autumn. Delayed seeding invites insect damage. Do not cover the seeds with soil as this often hampers germination; merely firm the soil after drilling. If the clover seed is broadcast in late autumn among the dying weeds and grasses on levees and roadsides, clover growth gradually thickens. When the clover is sown initially in the spring among the weeds, cut it back a year later to stimulate growth. Ladino clover vine may additionally be planted in spring in the same manner as sweet potato vine so as to ensure a full cover of clover by summertime.

Managing Ladino Clover: Clover does not choke out other vegetation, but gradually becomes dominant by growing so thickly as to prevent the germination and establishment of other weeds. Moreover, when trampled and cut, most weeds weaken but clover grows all the more vigorously. Failure to understand this and properly control the clover will lead to certain failure. At first, when the clover coexists with weeds, there may no cause for concern. But if, after the clover takes well and fluorishes, it is left alone, it becomes excessively luxuriant, leaving it open to attack by diseases such as leaf spot and the reemergence and eventual dominance of weeds again in five to six years. To maintain it over the years, clover requires the same meticulous care that one gives a lawn. Areas where perennial weeds such as sorrel and dandelion, twining plants such as bindweed, and cogon, bracken, and other herbs grow in abundance should be cut more frequently than other places, and wood ashes or coal ashes scattered.

The rate of lateral growth by clover is slow, so when starting the orchard, sow the seed from one end of the orchard to the other. With proper management, this clover cover will eliminate the need for weeding, and mowing will be incomparably easier than in an orchard overrun with weeds. I believe that, anyway one cares to look at the matter, ladino clover should be sown in citrus orchards as well as deciduous fruit orchards.

Alfalfa for Arid Land: Nothing surpasses ladino clover in dealing with weeds, but in warm regions where it tends to lose its vigor in the summer, and in cold, dry areas, mixed seeding with alfalfa is desirable. This works especially well on earthen levees, for example.

Alfalfa is very deep-rooted, sending roots down to depths of six feet or more. This makes it ideal for improving the deeper soil strata. A hardy perennial, it is of great practical value, being resistant to droughty and cold conditions as well as to high temperatures. When mixed with clover, alfalfa helps to eliminate other herbs and grasses. Wider use of this valuable legume should be made in Japan for soil improvement and as a feed and forage crop. Other legumes such as lupine (a summer crop) may also be used with good results.

Bur clover, useful in controlling spring weeds, withers in the summer but grows back again in the fall and suppresses winter weeds as well. A useful orchard cover crop, it also is valuable in the rotation as a crop preceding summer vegetables.

Morishima Acacia: Although Morishima acacia serves as a fertilizer tree, I would like to include it here because it plays a role also in association with ground cover cultivation. Up to about ten of these trees should be planted per quarter-acre among the fruit trees. A member of the pea family, this tree is effective in the following ways:

1)   rapid improvement of deep soil layers;
2)   can be used to form a shelterbelt, but may serve also as a windbreak when planted between fruit trees;
3)   serves as a shade tree during the summer in warm regions and protects

the soil from depletion;

4) effective in preventing the emergence of orchard pests, especially mites.

Nor is this all. The bark of the tree is rich in tannin and can be sold for a good price. In addition, the wood is excellent as a material for making desks and chairs, and the nectar of the flower serves as a source of honey.

No other evergreen tree of the pea family grows as quickly as Morishima acacia. It grows five feet or more in a year, creating a shelterbelt in just three to four years, and becoming about the size of a telephone pole in seven to eight years.

After five to six years of growth, I felled these and buried the trunks and tops in trenches within the orchard. Saplings do not take well, so it is better to plant the seed directly. All one has to do is scatter seed here and there throughout the orchard, and in six years or so, it becomes hard to tell from afar whether one is looking at a citrus grove or a forest.

Along with growing cover crops, I started early on to dig trenches and fill them with organic matter to speed up the process of soil enrichment. I tried using a variety of organic materials such as straw, hay, twigs and small branches, ferns, wood and bark chips, and lumber. After comparing the results, I found that hay, straw, and ferns, which I would have expected to be the least expensive, were in fact quite costly, while wood chips were not. The only problem was hauling this material in. As it turned out, the best material was lumber, which was relatively inexpensive, but this too was at times difficult to carry in. This is when I first decided to produce lumber right there in my orchard. Figuring that the easiest and most beneficial way was to return to the orchard what had been grown there, I tried planting various types of trees and found the best to be Morishima acacia.

Five or six years after planting acacias, an area of more than 100 square yards of what had been hard, lean soil about each tree had become soft and porous. This was far easier than blasting with dynamite and burying organic matter, and much more effective. In addition, when cut, each tree gave as much as a half-ton of high-quality organic material for burying. It was hard to feel enthusiastic about digging trenches when there was nothing to bury in them, but with organic material on hand, the trenches got dug.

Acacia Protects Natural Predators: I recommend the use of Morishima acacia even when replanting an old, rundown orchard. For example, in the case of a 40- to 50-year-old orchard, one could plant a large number of these acacia among the fruit trees and five or six years later fell all the fruit trees and acacias at once then replant the entire orchard with three- to four-year saplings. Not only would this be a far better method of replenishing the soil than running a bulldozer through the orchard and replanting, it would also rejuvenate the land.

The acacia grows constantly throughout the year, always sending out new shoots. These attract aphids and scales, which support a growing population of ladybugs. One important role of the acacia then is to serve as a protective tree for beneficial insects. Planting five or so acacia per quarter-acre keeps scales and mites down to a minimum. In addition to acacias, other trees that support populations of beneficial insects will certainly be developed in the future.

Some Basics on Setting Up a Ground Cover: I would like to go into a bit more detail here on the actual procedure for building up the soil with cover crops.

Once sown, a cover crop of clover remains hardy for about six to seven years, after which growth gradually slows. Although good management can extend the life of a stand of clover, by about ten years after the original planting the crop has declined to the point where weeds begin to reemerge. These weeds include primarily vines and climbing herbs such as bindweed and *kudzu*, and perennials such as the various sorrels. What happens is that those herbs resistant to clover survive and reestablish themselves.

Thus, perhaps ten years after the clover crop has been planted, the orchard is again overrun with weeds, but this need not present a problem as long as the weeds do not interfere with farming operations. In fact, when one stops to think about it, the soil tends to become imbalanced when a stand of one type of plant is grown year after year on the same land; the emergence and succession of different weeds is more natural and more conducive to soil enrichment and development.

I have no intention of insisting on a cover of clover; a weed cover will probably do just as well. The only concern I would have is that the weed growth become so thick as to be hard to cut back when necessary. If this happens, then one should sow clover seed again or switch to a cover of vegetable plants.

What should or should not be used as a cover crop for soil improvement depends largely on local conditions. All plants emerge for a reason. A succession of different herbs takes place over the years as the soil becomes richer. By sowing vegetable seeds of the same family as the weeds growing in the orchard, vegetable plants can eventually be made to replace the weeds.

These vegetables are fitting food for the young people living on a natural diet in the huts in my orchard. Large, hardy vegetables can be grown simply by scattering the seed of cruciferous vegetables in the fall, solanaceous vegetables in the spring, and leguminous vegetables in the early summer among the orchard weeds. I will come back to this later, but suffice it to say here that, in addition to being an effective means of controlling weeds, sowing vegetable seed among the weeds is also a powerful soil improvement technique.

One can understand the nature of the soil more quickly by examining the weeds growing in it than examining the soil itself. Weeds solve the problems of both the soil and the weeds. All I did was apply this belief to the restoration of barren soil and the trees and earth of an orchard tended for many years by scientific methods. It has taken me over thirty years and I admit it may not be much, but I have learned through natural farming how to naturally replenish the soil and what the natural form of a citrus tree is.

Soil Management: Soil improvement by natural farming takes a long time. Of course, with the large bulldozers around today, soil can be upgraded in a short time just by tearing everything up and throwing large amounts of coarse organic matter and organic fertilizer onto the land. Yet this requires tremendous outlays for equipment and materials.

Five to ten years are needed to build up six inches of topsoil through soil improvement by the cultivation of cover crops. To current economic perceptions, one disadvantage of natural farming methods is that they take too long. Perhaps these appear inferior in a world pressed for time, but if farmland is correctly understood as a legacy to be preserved for future generations, the general estimation of natural farming shall improve. Land that grows fertile over time without plowing, weeding, or chemical fertilizers represents not only an accumulation of labor and capital, but an increase in intangibles as well.

Physical improvement and the application of human effort alone have only a temporary effect. Natural farming makes use of the forces of living organisms to physically and chemically improve the soil, a process that goes hand-in-hand with the overall process of fruit growing. The beneficial effects of this approach ultimately show up in the longer lifetime of the fruit trees, which is perhaps two to three times that of fruit trees grown by scientific methods.

This is because, like the chickens, hogs, and cattle raised on artificial feed in cramped batteries and pens, fruit trees grown in artificially prepared soil with artificial fertilizers are inevitably weak, becoming either dwarfs or leggy and unable to live out their natural span of life.

Another reason has to do with the qualitative improvement in the soil. Obviously, scientific farming makes use of certain methods to improve poor soil. For example, if the soil is acidic, one applies lime or takes steps to prevent the excessive uptake of manganese or a deficiency in phosphates or magnesia. And if the soil is poorly aerated, root growth is poor, or insufficient zinc is present, a corrective is taken, such as replenishing the zinc. But if, on the other hand, the soil becomes alkaline, this leads again to a manganese and zinc deficiency, so even adjusting the soil acidity is no easy matter.

But there is far more to the quality of a soil than its acidity. An infinitude of factors and conditions—physical, chemical, biological—go into the overall assessment. Nor can one justifiably call a soil healthy or diseased as there are no criteria by which to judge whether a handful of soil contains the right number of certain microbes, the right amount of organic matter, and the right percentage of water and air.

Because it is convenient and for no other reason, we compare the merits of soil obtained through scientific farming with the soil of a natural orchard by looking at the amount of tree growth, the quantity and quality of harvested fruit, and whether the trees bear a full crop every year or only in alternate years. Even under such criteria, my thirty years of natural farming compare favorably with scientific farming in every respect. In fact, such comparison leaves the strong impression that scientific farming is more labor intensive and less efficient than natural farming.

I did not apply lime or any type of micronutrient, and yet noted no deficiencies. At no time did this ever become a problem. The constant change in the conditions of the cover crop within the orchard showed only that the soil changes constantly and that the fruit trees growing within that soil adapt constantly to such changes.

Disease and Insect Control

In nature, trees are constantly attacked and parasitized by insects and disease, but the widely accepted belief that unless the grower sprays his trees, they will succumb and die just does not hold in the wild.

Crops are more susceptible to such attack because they have been artificially improved, reducing their innate resistance, and the environment in which they are grown is unnatural. By selecting varieties of fruit trees closer to their natural ancestors and growing these properly, pesticides become unnecessary.

But certain insects and diseases present special problems in some types of fruit trees. Table 4.12 shows the degree of resistance various types of fruit trees have to disease and insect pests.

Table 4.12 Resistance of fruit trees to pests and disease.

Evergreen Fruit Trees	Major Pests	Control
Strong Resistance		
wax myrtle		
kumquat		
Moderate Resistance		
loquat	long-horned beetles, weevils	handpick
Japanese summer orange	scale insects	bags over fruit, natural enemies
Iyo orange, shaddock	scale insects	natural enemies
Weak Resistance		
Satsuma orange	scale insects, mites	natural enemies
sweet orange	long-horned beetles	handpick

Deciduous Fruit Trees	Major Diseases/Pests	Control
Strong Resistance		
plum, apricot, Chinese quince, Japanese apricot	black spot	companion planting
fig	wasps	
akebia, Chinese gooseberry, wild grape		
cherry		
persimmon (astringent)		
pomegrenate, jujube, oleaster, currant		
ginkgo, walnut		
Moderate Resistance		
nectarine	tree borers	companion planting
chestnut	tree borers	cleaning around tree
	chestnut gall wasp	resistant variety
persimmon (sweet)	persimmon fruit worm	cleaning around tree
Weak Resistance		
peach	tree borers	companion planting or bags over fruit
apple	tree borers	companion planting
pear	rust	resistant variety
grape	scarab beetles	lure and kill

The trees listed under "moderate" and "strong resistance" can be grown without the use of pesticides, provided some attention is given to a few specific diseases and pests. Clearly, the fruit grower should be thoroughly familiar with the characteristics and behavior of these important diseases and pests, and should take steps to prevent them from arising, such as selecting resistant varieties of trees.

Even so, the most difficult problem facing anyone growing fruits naturally will undoubtedly be the control of diseases and pests. There are a goodly number of fruit trees that can be grown without spraying. Although resistant types such as the peach, pear, grape, and Satsuma orange may not require the use of powerful pesticides, care must be taken with regard to certain pests. Let me give some of my observations regarding several of the most important.

Arrowhead Scale: Infestation of the Satsuma orange, Iyo orange, and shaddock by arrowhead scales has become so severe that an immediate stop to the spraying of citrus trees would be quite difficult, but damage by this pest can be overcome with natural predators and by correcting the form of the trees. Parasitic wasps and four or five different types of ladybugs emerged in my natural orchard.
In areas where these feast in large number on the scales, I have not sprayed and yet the trees have escaped serious damage. But even when these natural enemies are present, places where branches crisscross and are congested will sustain considerable damage unless the trees are pruned. No degree of spraying can succeed in effectively destroying arrowhead scales in trees with excessive branches and foliage.

Since the extent of disarray in the tree form and the degree of shade and sunlight have a large effect on the outbreak and persistence of scale infestation, I believe that the quickest and most effective solution is to protect the natural enemies that feed on this insect and to improve the microenvironment.

I find that spraying the trees with a machine oil emulsion in the winter or with a lime-sulfur mixture in the summer during the larval stage is effective. The latter application also destroys mites. There is no need to apply anything stronger than this. In fact, if you are not concerned about a minor loss in the tree's appearance, then you can certainly do without any spraying at all.

Mites: Up until about twenty to thirty years ago, a mixture of lime and sulfur was regarded as effective against fruit mites, and so growers in Japan sprayed their fruit trees with this twice each summer. As a result, mites never were an important pest.

Then after World War II, orchardists started applying powerful organophosphorus and organochlorine pesticides and were delighted that these destroyed all insect pests. But it was not long before many found that, no matter how often they sprayed, they were unable to prevent large outbreaks of mites from recurring.

Researchers offered a number of different explanations. One said that the mites had developed a resistance to the pesticides, a second that a different species of mite had emerged, and a third that the outbreaks resulted from the disappearance of natural enemies. One new pesticide was developed after another, but this only aggravated the problems of pest control and pesticide pollution.

Instead of speculating on the causes for these outbreaks, I prefer to concentrate on the fact that mite infestation at one time was not a problem. Many types of mites exist and each emerges under different conditions, but we can be sure of one thing: cultivation in the total absence of mites throughout the year is just not possible. Our goal should be to hold the damage they cause to a minimum, not total extermination.

Although the chances were always there for the emergence of mites in nearby trees, in shelterbelts, and in weeds, one never saw major outbreaks that killed trees and grasses. The causes for the recent infestations and the extensive damage to fruit trees lie not in the mites themselves but in human actions.

Mites are even more sensitive to microclimatic changes in the tree than are scales. When Morishima acacia is used as a windbreak or shade tree, depending on the amount of sunlight and breeze to which the tree is exposed, the number of mites and scales may drop dramatically or almost entirely vanish. Certainly part of the reason is that the Morishima acacia, which produces tannin, excretes a substance that repels insects, but the most direct cause of such rapid changes in population are changes in the microclimate.

The interplanting of evergreen trees with deciduous trees is also an effective preventive measure against infestation by these pests.

Given that not even the most rudimentary studies have been done on the effects of sunlight, ventilation, temperature, and humidity on mite infestation, how totally reckless it is then to try to control these with pesticides. What we have done is to spray potent pesticides without knowing anything about the relationships between the pesticides and the natural predators and beneficial fungi that feed on these mites. We have put the cart before the horse.

I do not expect this basic problem to be solved by the scientists. They are headed in some other direction, with such plans as the development of new pesticides that destroy pests at minimal harm to beneficial insects, and the construction of enormous ventilation and cooling towers.

If man had left the mite alone, it would never have become a major pest. I never had any problem with mites in the citrus trees in my orchard. Or if I had, the problem solved itself.

Cottony-Cushion Scale: At one time this was considered one of the three major citrus pests in Japan, but it disappeared naturally with the release more than thirty years ago of the vedalia, a kind of ladybug. After the war, a serious outbreak of this pest occurred in many orchards with the spraying of organophosphorus pesticides, and it became impossible to contain them. In my natural orchard, where I did not use strong pesticides, these continued as before to serve as the prey for several types of ladybugs, and so I saw almost no damage.

Red Wax Scale: This scale insect used to be another of the three major citrus pests and had to be destroyed by spraying a pine rosin mixture. In perhaps what was a stroke of good luck, at about the same time that applications of pine rosin compound were discontinued because of a wartime shortage of the rosin material, parasitic wasps emerged that preyed on this scale, making it no longer necessary to exterminate them.

But after the war, although the red wax scale was no longer much of a problem, farmers began to use a potent fluorine pesticide reputed to be effective against the scale. Severe outbreaks of the pest arose at once. Because this agent was highly toxic and even responsible for a number of local deaths, its use was later banned. Infestation by the scale declined almost immediately, demonstrating that the most intelligent way of controlling this particular pest was not to spray.

Other Insect Pests: There are an endless number of other fruit tree pests, such as aphids, tree borers, beetles that feed on grapevines, insects such as leaf rollers that attack leaves, and other insects such as springtails and grubs that feed on fruit. These become a problem in abandoned orchards in which no effort whatsoever is made to provide a good environment for the fruit trees or to improve their form. How much wiser it would be to keep the orchard clean and cope with insects while they overwinter in the larval stage. It is necessary, for example, to directly pick off and destroy the larvae of long-horned beetles that enter at the base of citrus and chestnut trees. These tend to attack weakened trees and trees in neglected orchards.

Now I would like to take a look at two pests of foreign origin that may become a problem in Japan.

Mediterranean Fruit Fly and Codling Moth: With the current "liberalization" of international fruit trade, we have recently been seeing unrestricted imports into Japan of oranges and grapefruit from Europe and Africa as well as apples from northern countries. It seems almost inevitable that with these fruit we shall soon see the entry of the Mediterranean fruit fly and the codling moth, pests capable of becoming a far greater headache to the Japanese farmer than the fruit imports that he so fears.

The maggots of the Mediterranean fruit fly attack not only Japanese citrus trees, pears, peaches, apples, and melons, but also vegetables such as eggplants, tomatoes, and cucumbers—indeed, all major fruit and vegetable crops, while the codling moth ravages apples, pears, and other fruit of the rose family. Extermination of these will be difficult if not impossible; once they have entered Japan, they may very well cause incredible damage. It is no exaggeration to say that one vital mission of plant quarantine operations at Japanese customs is to prevent the entry of these pests into Japan. That these operations have been successful thus far is a testimony to their thoroughness.

The importation of fruits and vegetables grown along the Mediterranean Coast in Europe and in Africa, and apples from Manchuria and other northern countries is strictly banned at customs to prevent the entry of these two pests. Until now, strict laws have been enforced forbidding the entry of even one of these fruits from these areas, but with the open and unrestricted importation of fruits in the future, the arrival of these pests on Japanese soil is almost inevitable. The consequences are almost certain to be far greater than a mere lightening in the duties of plant inspection officials.

The larval worms and maggots of these pests bury deep into the fruit where outside spraying and fumigation has no effect. The only possibility is physical measures such as cold storage, but these are not likely to be effective without damaging the quality of the fruit. The spread of these pests in Japanese fields and orchards will be a strong blow to Japanese farmers and become an immense burden.

I would simply like to warn that the free movement of fruit may satisfy the fleeting desires of people, but the price we will have to pay will be enormous. This is exactly what happened recently in the United States with the Mediterranean fruit fly.

The Argument against Pruning

Pruning is the most difficult of the skills practiced by fruit growers. Growers prune their fruit trees to shape them and adjust the vigor of the tree so as to maintain a balance between tree growth and setting of the fruit. Trees are also pruned to increase the yield and quality of harvested fruit and to facilitate orchard management and operations such as pesticide spraying, tillage, weeding, and fertilization.

No Basic Method: Although pruning is of utmost importance in fruit growing, no single basic method is practiced. In addition, it is often difficult to know how much pruning is enough. The grower usually has no choice but to switch back and forth among a variety of different pruning methods as the immediate circumstances seem to require. With all the local variance in methods and opinions and perhaps also because of the many years of experience and experimentation that have been devoted to it, pruning has done more to confuse fruit growers than any other aspect of orcharding. One question that deserves to be asked then is whether pruning really is a necessary part of fruit growing in the first place. Let us examine the motives and reasoning that led farmers to start pruning.

If pruning is discontinued on a fruit tree, the form of the tree becomes confused, the primary scaffold branches entangle, and the foliage grows dense, complicating all orchard management. Heavy spraying of pesticides becomes ineffective. As the tree grows older, the branches become ridiculously long, crossing with the branches of neighboring trees. Sunlight ceases to penetrate the canopy to the lower branches, which weaken as a result. Ventilation is poor, encouraging infestation by disease and insects. Dead and dying branches abound. Fruit ends up by forming only at the surface of the tree. It is quite possible that, having observed this occurring in orchards, growers came to regard pruning as absolutely essential.

Another motive for pruning has to do with the reciprocal relationship between tree growth and fruit bearing effects. When tree growth is too vigorous, the tree bears little fruit; on the other hand, when a tree bears too much fruit, growth declines. Thus, in years when a poor crop is anticipated, one prunes to promote fruit setting and the bearing of high-quality fruit. But in years when a tree looks as if it will bear too heavily, then it must be pruned to increase vigor and growth. The grower has to constantly adjust tree growth and fruit formation to prevent the tree from growing into a tangled and disorderly shape and bearing a full crop only in alternate years. This certainly seems to justify the development of intricate and complicated pruning techniques.

But if, instead of being neglected or abandoned, the tree is left to grow in its natural form, this is altogether a different matter. Yet no one has ever really seen a totally natural fruit tree or given any thought as to what a natural fruit tree is. Nature is a world simple and close at hand, yet at the same time distant and inaccessible. Although man cannot know what a truly natural tree is, he can search for the shape of a tree that comes closest to its natural form.

When a tree is left to grow by itself under natural circumstances, how likely are its primary scaffold branches to crisscross and its smaller branches and foliage to crowd each other? Would it be reasonable to expect the tree to put out leaves and branches not touched by the sun? Would it seem normal for lower and inner branches to die back? For fruit to form only at the ends of branches? This is not the form that a natural tree takes, but one most commonly seen in trees that have been pruned haphazardly then abandoned.

Take a look at the pines and cedars that grow in natural forests. The trunks of these trees never branch or twist as long as they are not cut or harmed. The branches on the right and left sides of the tree do not run up against each other or cross; there are no dense lower branches that die back; upper and lower branches do not grow so close that sunlight cannot reach some of the leaves. No matter how small the plant or large the tree, every leaf, every shoot and branch grows out from the stalk or trunk in an orderly and regular arrangement. No part of the plant is in disarray or confusion.

For instance, in a given plant, leaves always grow either alternately or oppositely. The direction and even the angle at which a leaf grows is always the same; never is there even the slightest deviation. If the angle between one leaf on a fruit tree branch and the next leaf is 72 degrees, then the next leaf and all the other leaves too will emerge at respective angles of 72 degrees. The arrangement of the leaves on a plant always and unerringly obeys a fixed law called phyllotaxy. Thus, the sixth leaf on the branches of peach, persimmon, mandarin orange, orange, and cherry trees is always located directly above the first leaf, and the eleventh leaf is always directly above the sixth leaf. When the distance along the branch between two consecutive buds is one inch, then the distance from one leaf to the next leaf directly above it is always five inches. Two leaves will not overlap, or two branches emerge, within any five-inch length along the branch.

The direction, angle, and divergence of a shoot or branch is regular and orderly. Never does one branch cross over another; lower and upper branches maintain the same distance over their entire length, never overlapping. This is why the branches and leaves of natural plants all receive equal ventilation and sunlight. Not a single wasted leaf, not a single branch lacking; that is the true form of a plant.

All this is abundantly clear when one looks carefully at a mountain pine. The central trunk rises straight and true, putting out branches at equal vertical spacings in a radial arrangement. One can clearly make out the chronology of branch emergence, the spacing and angle of the branches being also regular and orderly. Never does one branch grow too long or cross with another branch.

In the case of bamboo, the emergence of a branch or leaf follows a fixed law for that type of bamboo. Likewise, cryptomeria, Japanese cypress, the camphor tree, camellia, Japanese maple, and all other trees observe the phyllotaxy and divergence specific for that species.

What happens if we simply let fruit trees and mountain pines grow to their full size under natural conditions? The very goal at which the gardener or fruit grower aims through pruning is attained naturally by the tree without the intertwining, clustering, or dying back of branches. Had the persimmon, the peach, and the citrus tree been left to grow of their own accord, it would never have been necessary to cut the trunks with a saw or lop off branches to control erratic growth.

Just as no one is so foolish as to strike his left hand with his right, no persimmon or chestnut tree has branches on the right that compete with those on the left and eventually have to be cut back because they grow too long. A branch on the east side of the tree does not wander over to the south side cutting off light. And what tree grows inner branches only to have them die off because they receive no light? There is something strange about having to prune a tree in order for it to bear a full crop of fruit each year, or having to balance growth of the tree with fruit formation.

A pine tree produces pine cones, but if someone were to prune the pine to promote growth or retard fruit formation, the result would be quite curious. A pine tree grows just fine under natural conditions and requires no pruning. In the same way, if a fruit tree is grown under natural conditions right from the start, there never should be any need for pruning.

Misconceptions about the Natural Form: Orchardists have never tried growing fruit trees in their natural form. To begin with, most have never even given any thought as to what the natural form is. Of course, pomologists will deny this, saying that they are working with the natural form of fruit trees and looking for ways to improve on this. But it is clear that they have not really looked in earnest at the natural form. Not a single book or report has been published which discusses pruning based on such basic factors as the phyllotaxy of a citrus tree, or which explains that a divergence of so much gives such-and-such a natural form with primary and secondary scaffold branch angles of X degrees.

Many have a vague idea of the natural form as something akin to the shape of a neglected tree. But there is a world of difference between the two. In a sense, the true natural form of a tree may be unknowable to man. People will say that a pine tree should look like this, and a cypress or cedar like that, but knowing the true form of a pine tree is not all that easy. It is all too common for people to ask whether a low, twisted pine on the seashore is the natural form, and to become perplexed as to whether a lone cryptomeria standing tall in a meadow with alternate branches drooping downward in all directions is the natural form for this tree or whether the branches should be inclined upward at an angle of 50 degrees and ranged radially about the trunk like a mountain pine.

Like the camphor tree transplanted into a garden, the flowering camellia buffeted by high winds on an exposed coast, the Japanese maple perched above a waterfall, and other trees scratched, pecked, and attacked by bird, beast, and insect, plants grow under an incredible diversity of conditions. And so it is with fruit trees. To go off in pursuit of *the* natural form of the peach tree, or the citrus tree, or the grapevine is to miss the point altogether.

Scientists say that the natural form of a citrus tree is hemispherical with several primary scaffold branches extending out like the ribs of a fan at an angle of from 40 to 70 degrees, but in truth no one knows whether the true form of a citrus tree is that of a large, upright tree or a low bush.

It is not known whether this grows like a cryptomeria with one tall central trunk, in the manner of a camellia or Japanese maple, or round like paperbush. Persimmon, chestnut, apple, and grape too are pruned by growers who have not the slightest idea of what the natural forms for these are.

Fruit growers have never really been too concerned with the natural form of a tree and are not likely to become so in the future. This is not without reason. In a system of cultivation based largely on activities such as weeding, tillage, fertilization, and disease and pest control, the ideal form of a tree is the form best suited to these various human operations and to harvesting. Thus it is not after the natural form that gardeners and growers seek, but after a shape artificially pruned and trained to the convenience and benefit of the grower. But is it really in the best interests of the farmer to rashly prune his trees without having any idea of what the natural form is or the slightest inkling of the powers and subtlety of nature?

Fruit growers have more or less decided that, if one considers such operations as harvesting of the fruit, pesticide spraying, and fumigation, the ideal form of citrus trees grown in a hillside orchard is a round, flat-topped shape measuring at most about 9 feet high and 14 feet in diameter. To improve fruit production, the grower also thins the trees and does some cutting back here and there with the pruning shears. Deciding that a grapevine should be trained on a single main trunk or on a trunk and two laterals, he prunes all other branches. He takes a saw to the leader on a peach sapling, saying that a "natural" open-center shape with a scaffold of three strong branches is best. In pear trees, the two or three main branches are set at angles of 40 or 50 degrees or drawn out horizontally, and all the other smaller branches pruned during the winter. A modified leader system is said to be best for persimmon trees, so leader growth is checked by nipping the tip, and many branches either cut back or removed altogether.

Is Pruning Really Necessary?:   I would like to turn back now and look at why pruning is necessary, why growers must remove so many branches and leaves. We are told that pruning is essential because lower branches get in the way during tillage, weeding, and fertilization, but what happens when we eliminate the operations of weeding and tillage? We no longer have to worry about the convenience of the tree shape for any operations other than fruit-picking. Pruning has always been just something that fruit growers felt they had to do to bring the shape of of the tree in line with the form they visualized as ideal for all other orchard operations.

Pruning is necessary for another reason as well. Like the transplanted mountain pine to the top of which the gardener takes his shears, once pruned, a tree cannot be left untended. The branches of a tree growing naturally never cross or entangle, but once even the smallest part of a new shoot is damaged, that wound becomes a source of confusion that follows the tree for life.

As long as the shoots on a tree emerge in an orderly fashion according to the natural law for that species, guarding the correct angle front and back, left and right, there is no crossing or entangling of the branches. But if the tip of just one of these branches is pinched off, several adventitious buds emerge from the wound and grow into branches. These superfluous branches crowd and become entangled with other branches, bending, twisting, and spreading confusion as they grow.

Because even lightly pinching new buds on a pine seedling totally alters the shape of the emerging branches, the young tree can be trained into a garden pine or even a *bonsai*. But although the first pruning can make a *bonsai* of a pine, once a *bonsai*, the pine can never be restored to a full-size tree.

The gardener prunes the young shoots of a pine planted in the garden and the second year several suckers grow out from each of these wounds. Again he cuts the tips of these and by about the third year, the branches of the pine become entangled and crooked, taking on an incredibly complex shape. Since this is precisely what gives it its value as a garden tree, the gardener delights in topping confusion with more confusion.

Once the pruning shears have been taken to the tree and branches emerge in complicated shapes, the tree can no longer be left alone. Unless it is carefully tended each year and each branch meticulously trained and pruned, the branches entangle, causing some to weaken and die. Seen from a distance, there may not seem to be much difference between a garden pine and a mountain pine, but on closer inspection one can see that the confused and complicated shape of the garden pine has been artificially modified to allow sunlight to fall on each branch and leaf, while the natural pine achieves the same goal without any help from man.

The question of whether a fruit tree should have a natural form or an artificial form is directly analogous to the question of which is preferable, a natural pine or a garden pine.

A fruit tree sapling is first dug up and the roots trimmed, then the stem is cut back to a length of one or two feet and the sapling planted. This first pruning operation alone robs the tree of its natural form. The sapling begins to put out buds and suckers in a complex and confused manner that requires the fruit grower to be always at the ready with his pruning shears.

People will stand in front of a citrus tree and, saying that these branches here are growing so closely that they are shutting off sunlight, casually make a few quick cuts with the shears. But they never stop to consider the enormous impact this has on the tree. Because of this single pruning, the grower will have to continue pruning the tree for the rest of its lifetime.

Just by nipping one bud at the tip of a sapling, what should have grown into a straight pine with one trunk develops instead into a complex tree with several leaders; a persimmon comes to resemble a chestnut and a chestnut takes on the form of a peach tree. If the branches of a pear tree are made to crawl along a netlike trellis seven feet off the ground, then pruning is absolutely indispensable. But if the tree is allowed to grow up straight and tall like a cedar, initial pruning is no longer necessary. Grapevines are grown over metal wires, but they can also be grown upright like a willow tree with pendant branches. How the first leader is trained determines the shape of the vine and the method of pruning.

Even slight training of the branches or pruning when the tree is young has an enormous effect on the later growth and shape. When left to grow naturally from the start, little pruning will be needed later on, but if the natural shape of the tree is altered, a great deal of intricate pruning becomes necessary. Training the branches at the start into a shape close to the natural form of the tree will make the pruning shears unnecessary.

If you draw a mental picture of the natural form of a tree and make every effort to protect the tree from the local environment, then it will thrive, putting out good fruit year after year. Pruning only creates a need for more pruning, but if the grower realizes that trees not in need of pruning also exist in this world and is determined to grow such trees, they will bear fruit without pruning. How much wiser and easier it is to limit oneself to minimal corrective pruning aimed only at bringing the tree closer to its natural form rather than practicing a method of fruit growing that requires extensive pruning each and every year.

The Natural Form of a Fruit Tree

The art of pruning fruit trees is the most advanced skill in orcharding, and is even said to separate the good farmer from the bad farmer. Although I have, as I advocated in the preceding section, grown fruit trees without pruning, I found this very difficult going at first because I did not know what the natural forms of the different types of fruit trees were. To learn of these forms, I began observing various plants and fruit trees.

The natural forms shown from time to time in journals on fruit growing are not at all what they are made out to be. These are just abandoned trees of confused shape that have been left to grow untended after having been initially pruned and otherwise cared for. It was relatively easy to determine that the natural form of most deciduous fruit trees is a central leader system, but I had a lot of trouble determining the natural form of citrus trees, and especially the Satsuma orange.

I first tried applying the methods of natural farming to an established grove of Satsuma orange trees with a couple of hundred trees to the acre. Trees at the time were trimmed in the shape of a wineglass and the height held to about six or seven feet. Because I simply discontinued pruning, letting these trees grow untended, large numbers of scaffold branches and laterals grew out at once. All of a sudden, these began crisscrossing, doubling back, and growing in strange, twisting shapes. Places where the branches and leaves grew tangled became the site of disease and drew insects. One dying branch caused other branches to wither and die. The confused shape of the tree resulted in irregular fruit formation. Fruit grew either too far apart or too close together and the tree produced a full crop only every other year. This experience forced even me to admit that abandoning the trees to their own devices was a sure path to ruin.

To correct these gross disorders I then tried the reverse: heavy pruning and thinning. I left only several of the rising suckers remaining. Yet, because four or five primary scaffold branches were still too many, there was too little space left between adjoining branches and there may also have been too many laterals. In any case, growth at the center of the trees was poor and the inner branches gradually withered, causing a sharp drop in fruit production in the interior portions of the trees. Well, this experience taught me that abandoning the trees was the wrong way to approach their natural form.

Following the end of the war, specialists began advocating a natural, open-centered system. This consisted of removing scaffold branches at the center of the tree, but leaving several scaffolds projecting outward at angles of about 42 degrees, with two or three laterals growing from each scaffold branch. Since abandoned wineglass-shaped trees on which the rising scaffold branches had been thinned closely resembled this natural open-centered form, I gave some thought to moving in this direction.

Yet my ultimate goal remained to practice natural farming and so the question I faced was how to make it possible not to prune naturally. I thought that pruning would not be needed if the tree assumed its natural form. As I went from a wineglass shape to a neglected tree form to corrective pruning, I began to ask what shape was truly the natural form of the citrus tree. This led to my doubts about existing views.

The natural forms shown in illustrations in technical books and journals all showed hemispherical shapes with several scaffold branches meandering upwards. But my own unpleasant experiences had taught me all too clearly that these so-called natural forms were not true natural forms at all, but the shapes of abandoned trees. A natural tree does not wilt of its own accord. This is the result of some unnatural element. For reasons I will get into later, in my search for the natural form, I was to sacrifice another 400 citrus trees—about half of those in my care.

If a tree dies when left unpruned, this can be explained scientifically as the result of overcrowding between adjoining scaffold branches and laterals, which implies a need to know the proper spacing of these branches. These spacings can eventually be determined, or so it is thought, through experimentation and the application of human knowledge, and the proper number of inches calculated for given conditions. But never do we get a definitive spacing that is okay for all situations. A different result is obtained for wineglass-shaped trees, trees with natural open-center shapes and every other shape. The conclusion that each has its merits and demerits leaves the door open to continuous change with each passing age. This is the way of scientific agriculture.

If one takes the viewpoint of natural farming, however, there is no reason why the branches and foliage of trees having a natural form should ever become tangled and wither. If the tree has a natural form, then there should be no need for research on the desirable number of scaffold branches, the number and angle of the lateral branches, and the proper spacing between adjoining branches. Nature knows the answers and can take care of these matters quite well by itself.

Everything is resolved then if we let the tree adopt its natural form through natural farming. The only problem that remains is how to induce the tree to grow in its natural form. Simply abandoning it leads only to failure. Before being abandoned, my citrus trees had been trained and pruned into a wineglass shape. The trees had an unnatural form from the moment they were transplanted as saplings. This is why, when left unpruned, they did not return to a natural form but became instead increasingly deformed.

Obviously, the proper way to grow a citrus tree having a natural form would be to plant the seed directly in the orchard. But the seed itself, if I may press the point, is no longer truly natural. This is the product of extensive cross-breeding between different varieties of artificially cultivated citrus trees; if allowed to grow to maturity, the tree either reverts to an ancestral form or produces inferior hybrid fruit. Direct planting of the seed, therefore, is not a practical option for fruit production. Yet this is very helpful in gaining an idea of the natural form of the citrus tree.

I planted citrus seed and observed the trees growing from these. At the same time, I allowed a large number of various types of citrus trees to go unpruned. From these results, I was able to divine with considerable certainty the natural form of a citrus tree.

When I reported my findings at a meeting of the Ehime Prefecture Fruit Growers Association, stating that the natural form of the citrus tree is not what it had been thought to be, but a central leader type form, this created a stir among several specialists present, but was laughed off as just so much nonsense by farmers.

The natural form of a citrus tree is constant and unchanging in natural farming and permits pruning to be dispensed with. Whatever new pruning techniques may arise in the future, knowing the true natural form of citrus and other fruit trees, and how to train a tree to its natural form can never be a disadvantage.

For example, even when performing surgery on a tree in a mechanized orchard, it makes more sense to work on a tree trained on a single stem than to allow the tree to grow as much as it can and later cut it with a saw. The closer the form of the tree to nature, the more reasonable on all counts. When for purely human reasons there is absolutely no alternative, then the wisest choice is to adopt a form that is basically natural but makes some compromises.

The very first thing that one must do when preparing to grow a type of fruit tree by the methods of natural farming is to know the natural form for that fruit tree. In the case of Satsuma orange trees, the scaffold branches do not grow all that straight because the tree is not very vigorous. As a result, there is a great deal of individual variation between the trees, making it most difficult to discern the natural form. Few trees are as sensitive as these in the way they take on myriad different forms upon the slightest human tampering or injury. To determine the natural form of citrus trees, I chose to look at a cross-section of hardier and more vigorous citrus varieties than the Satsuma orange. The summer orange and the shaddock were especially useful in this regard. Both are clearly of the central leader type.

To determine the natural forms of persimmon, chestnut, pear, peach, and other trees, it was necessary to look at these from a broad perspective. Of course, each is grown in many different forms, but all are basically central leader type trees.

Their differences in form arise primarily from the differing number, angle, and directions of the scaffold branches that grow from the central leader. In form, they resemble forest trees such as the cryptomeria, Japanese cypress, pine, and live oak. People have merely been misled by the various forms that these fruit trees have taken after being disturbed by their environment and human intervention.

Examples of Natural Forms

Early-ripening Satsuma orange:	low, pyramidal form
Late-ripening Satsuma orange:	tallish, cypress-like conical form
summer orange, shaddock, persimmon, chestnut, pear, apple, loquat:	tall, cedar-like conical form

Fig. 4.7 Forms of fruit trees.

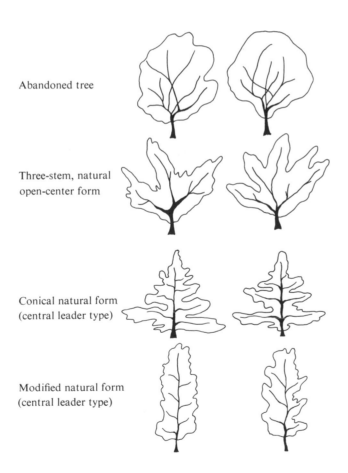

Abandoned tree

Three-stem, natural open-center form

Conical natural form (central leader type)

Modified natural form (central leader type)

Attaining the Natural Form: The shaddock and summer orange tend to have an upright central trunk and a height greater than the spread. These can even resemble a cedar in aspect, whereas the Satsuma orange generally has an irregular flattened shape or a hemispherical shape. This basic central leader type conical shape can occur in an essentially infinite number of variations depending on the type of tree and the cultivation conditions. The fact that few mandarin orange trees grown in their natural form take on a central leader type form, but adopt instead various modifications indicates that these trees have weak terminal bud dominance and tend to develop an open crown. They are frutescent, having several scaffold branches extending with equal vigor that produce a confused form. It is clear then that while many types of trees do fully retain their innate character, other trees have natural forms that are easily upset during cultivation.

Fig. 4.8   Forms of the mandarin orange tree.

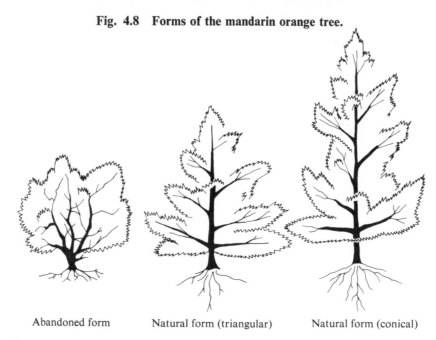

Abandoned form   Natural form (triangular)   Natural form (conical)

Natural Form in Fruit Tree Cultivation: I adopt the natural form of a tree as the model for the basic tree shape in citrus cultivation. Even when something causes . a tree to take on a shape that deviates from the natural form or adapts to the local environment, any pruning and training done should attempt to return the tree to its natural form. There are several reasons for this.

1.   The natural form permits tree growth and development best suited to the cultivation conditions and environment. No branch or leaf is wasted. This form enables maximum growth and maximum exposure to sunlight, resulting in maximum yields. On the other hand, an unnatural form created artificially upsets the innate efficiency of the tree. This reduces the tree's natural powers and commits the grower to unending labors.

2.   The natural form consists of an erect central trunk, causing little entanglement with neighboring trees or crowding of branches and foliage. The amount of pruning required gradually decreases and little disease or pest damage arises,

217

Fig. 4.9 The natural forms of deciduous fruit trees.

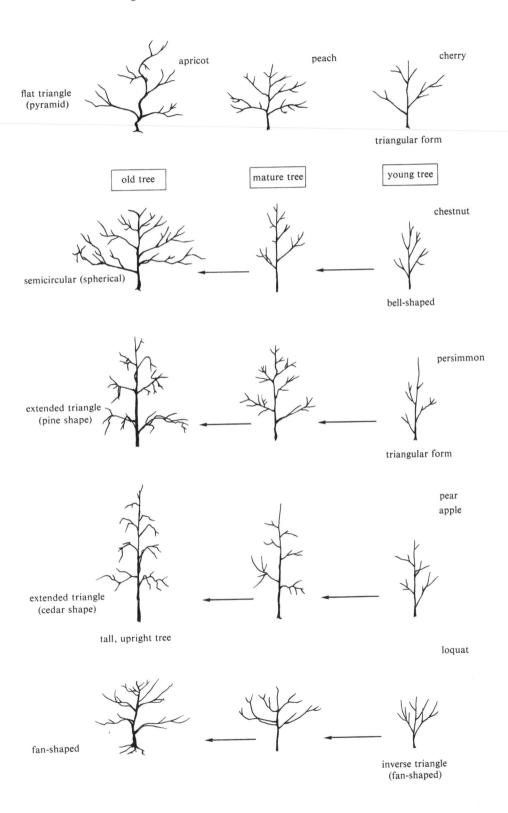

apricot

peach

cherry

flat triangle
(pyramid)

triangular form

old tree

mature tree

young tree

chestnut

semicircular (spherical)

bell-shaped

persimmon

extended triangle
(pine shape)

triangular form

pear
apple

extended triangle
(cedar shape)

tall, upright tree

loquat

fan-shaped

inverse triangle
(fan-shaped)

necessitating only a minimum of care. However, in natural open-center systems formed by thinning the scaffold branches growing at the center of the tree, the remaining scaffold branches open up at the top of the tree and soon entangle with adjacent trees. In addition, secondary scaffold branches and laterals growing from several primary scaffold branches oriented at unnatural angles (such as in three-stem systems) also crisscross and entangle. This increases the amount of pruning that has to be done after the tree has matured.

3.   In conical central leader type systems, oblique sunlight penetrates into the interior of the tree, but in open-center systems, the crown of the tree extends outward in the shape of an inverse triangle that reduces the penetration of sunlight to the base and interior of the tree, inviting the withering of branches and attack by disease and pests. Thus, expanding the shape of the tree results in lower rather than higher yields.

4.   The natural form provides the best distribution and supply of nutrients to the scaffold branches and laterals. In addition, the external shape is balanced and a good harmony exists between tree growth and fruit production, giving a full fruit harvest each year.

5.   The root system of a tree having a natural form closely resembles the shape of the aboveground portion of the tree. A deep root system makes for a healthy tree resistant to external conditions.

Problems with the Natural Form: Although having many advantages, the natural form is not without its share of problems in fruit growing.

1.   The natural forms of young grapevines and persimmon, pear, and apple trees have low branch, leaf, and fruit densities, and thus produce small yields. This can be resolved by discreet pruning to increase the density of fruit and branch formation.

2.   Fruit trees with a central leader system grow to a good height and may be expected to pose climbing problems when it comes time to pick the fruit. While this is true when the tree is still young, as it matures, scaffold branches grow out from the leader at an angle of about 20 degrees to the horizontal in a regular, spiraling arrangement that make it easier to climb. In tall trees such as persimmon, pear, apple, and loquat, this forms a framework that can be climbed much like a spiral stairway.

3.   Creating a pure natural form is not easy, and the tree may deviate from this if adequate attention is not given to protective management at the seedling stage. This can be corrected in part by giving the tree a modified central leader form. To achieve an ideal natural form, the tree must be grown directly from seed or a rootstock tree grown in a planting bed and field-grafted.

4.   Enabling the seedling to put out a vigorous, upright leader is the key to successfully achieving a natural form. The grower must observe where and at what angle primary and secondary scaffold branches emerge, and remove any unnatural branches. Normally, after five or six years, when the saplings have reached six to ten feet in height, there should be perhaps five or six secondary scaffold branches extending out in a spiral pattern at intervals of about six to

twelve inches such that the sixth secondary scaffold branch overlaps vertically with the first. Primary scaffold branches should emerge from the central trunk at an angle of 40 degrees with the horizontal and extend outwards at an angle of about 20 degrees. Once the basic shape of the tree is set, the need for training and pruning diminishes.

5.   The tree may depart from a natural form and take on an open-center form if the central leader becomes inclined, the tip of the leader is weak, or the tree sustains an injury. There should be no problem though, as long as the grower keeps a mental image of a pure natural form and prunes and trains the tree to approach as closely as possible to that form. A tree that has become fully shaped while young will not need heavy pruning when mature. However, if left to grow untended when young, the tree may require considerable thinning and pruning each year and may even need major surgical reconstruction when fully grown. Considering the many years of toil and the losses that may otherwise ensue, it is certainly preferable to choose to do some formative pruning early on.

Armed with confidence in my understanding of the natural form of these fruit trees, I saw clearly the basic approach I had to take in fruit cultivation. Later, when I extended my orchard by planting a new slope with fruit trees, I began with the goal of achieving this natural form in all the trees. But because this involved planting several thousand additional trees almost single-handedly, I was unable to establish the natural form I had intended. Still, these were closer to the natural form than the previous trees and thus required far less pruning. In fact, I managed to get by with almost no pruning at all.

Here then are the greatest merits of using the natural form in fruit growing.

1.   Attaining the natural form through early formative pruning minimizes waste and labor on all counts, and enables high fruit production.

2.   A deep-rooted tree adapted to the local environment and maintaining a good balance between the aboveground portion of the tree and the root system grows rapidly, is healthy, cold-hardy, frost and drought resistant, and stands up well to natural disasters.

3.   The absence of unnecessary branches minimizes the amount of pruning. Good light penetration and ventilation reduce the possibilities of bearing a full crop only in alternate years and of attack by disease or insects.

4.   Should the form of the tree have to be changed to adapt to local topography or mechanized practices, pruning back can be done smoothly and without undue difficulty.

5.   The pruning techniques used in fruit growing tend to change with the times, but the natural form of a tree remains always the same. Use of the natural form is the best approach possible for stable, labor-saving, high-yield fruit cultivation. Success is especially easy with trees such as the persimmon, chestnut, apple, pear, and loquat, which can readily be trained to a natural form. Considerable success can also be had with vines such as Chinese gooseberry and grape.

Conclusion

Fruit growing today relies heavily on practices such as weeding, tillage, fertilization, and pruning. I have described above the basics of an alternative way of orcharding, a natural method founded on a return to nature that allows a young sapling to grow into a tree with a close-to-natural form. Weeding is not used; instead, the living orchard soil is preserved and actively enriched. The fruit trees grow up sturdy and healthy without fertilizers, orderly and beautiful without pruning. These principles of no weeding, no fertilization, and no pruning cannot be achieved independently; they are closely and inextricably tied to each other.

Soil management techniques such as green manure cultivation and sod cultivation that eliminate the necessity of weeding and tillage at the same time make fertilizer-free cultivation possible, but attempts to suddenly do away with fertilization or weeding are not likely to succeed.

With pest and disease control it is the same; the best method of control is no control at all. In principle, disease and pest damage do not exist. If a farming system without weeding, fertilization, or pruning is established, crop damage by diseases and pests will gradually decline.

One reads in the news these days of how rangers are spraying mountain forests with fertilizers and herbicides to stimulate growth, but this is likely to have the undesirable effect of inducing disease and pest damage and necessitating even more complex spraying and fertilization operations. Plants grown without fertilizers in rich soil have strong, healthy roots and tops that are resistant to disease. Weeding, fertilization, and pruning confuse the soil and the tree, and reduce its disease resistance. The result is poor ventilation, branches and leaves not reached by sunlight, and infestation by disease microbes and insects. It is this that has created a need for disease and pest control. Today, by spraying their orchards with pesticides, fruit growers increase disease and pest damage; by pruning, they create strange, misshapen trees; and by applying fertilizer, they promote nutrient deficiencies.

The ultimate decision in favor of scientific farming or natural farming depends on what it is that man seeks.

4. Vegetables

Natural Rotation of Vegetables

Ideally, crops should be left in nature's care and allowed to grow in an almost natural state rather than being grown under artificial conditions by man solely for his own purposes. Crops know where, when, and how to grow. By sowing a mixture of many field crops, allowing them to grow naturally, and observing which thrive and which do not, one finds that, when grown in the hands of nature, crops superior to what would normally be imagined can be obtained.

For instance, when the seeds of different grains and vegetables are mixed together and scattered over growing weeds and clover, some vanish and some survive. A few even fluorish. These crops flower and set seed; the seed drops to the ground and is buried in the soil where the seed casing decomposes and the seed germinates. The seedling grows, competing with or being assisted by other plants. This process of growth is an amazing natural drama that appears at first disordered, but is eminently rational and orderly. There is much to be learned from the wondrous hand of nature.

Although this method of mixed, semi-wild cultivation may appear reckless at first, it more than suffices for the small family garden or for vegetable gardening on barren land by those who seek to live self-sufficiently.

However, for permanent cultivation on large acreages, this type of natural cultivation must be carried a step further. Systematic rotation schemes must be set up and cultivation planned and carried out in accordance with these. The natural crop rotation diagrams in Figs. 4.2 and 4.3 at the beginning of this chapter are intended to serve as a guide. The basic aim of such a system, which borrows some ideas from natural cropping, is to permanently preserve nature. But of course, it falls short of nature and must be complemented by whatever means and resources are called for under the circumstances.

The rotations in these diagrams provide for soil enrichment with leguminous green manure plants, the replenishment of organic materials with gramineous plants, deep working and conditioning of the soil with root vegetables, and reduced disease and pest damage as well as cooperative effects through the segregation of key vegetables of the Potato, Gourd, and Mustard families, and also the intermittent mixed planting of vegetables and herbs of the Lily, Mint, Carrot, and Composite families. This I have made the basis for a natural rotation system.

Although not all of the rotation schemes in the diagrams are ideal from the standpoint of nature, they are designed to move away from existing short-term rotation schemes that primarily benefit man and toward systems that benefit the earth. Their ultimate aim is to do away with tilling, fertilizers, pesticide application, and weeding.

No tilling: This consists typically of ridging the field at intervals of 3 to 6 feet or digging drainage channels every 13 to 16 feet the first year, then either not plowing the next year or, at most, shallow plowing followed by seeding and rotary tillage.

No fertilizer: Leguminous green manure is grown as a basic crop each year and a mixture of coated crop seeds sown. If direct sowing is not possible, seedlings are transplanted. In addition, the land is enriched without plowing or tilling by planting root crops throughout.

No weeding: The second crop is either seeded over the maturing first crop or transplanted prior to harvest so as to minimize the period during which the field is left fallow. The straw and leaves from the crops just harvested are used as a mulch to retard weed emergence while the second crop in the rotation is still very young.

No pesticides: Of course, one can also make use of plants that prevent or inhibit the emergence of diseases and insect pests, but true non-control can be achieved when all types of insects and microorganisms are present.

An effective natural crop rotation scheme therefore allows plants of all kinds to coexist, enables the soil to enrich itself, and provides soil microbes with a good environment in which to thrive.

Semi-Wild Cultivation of Vegetables

Producing and shipping naturally grown vegetables out to market for sale as natural food is far from easy. Problems exist both with the producer and with the market and consumer. However, as long as the farmer adheres closely to the natural vegetable rotation scheme and pays attention to the following points, productivity will be high.

A Natural Way of Growing Garden Vegetables: Vegetables grown for home consumption are most likely to be raised either for a five- or six-member family on a small plot of perhaps 100 square yards next to the house, or in a larger field. When grown in a small garden plot, all that is involved is growing the right crop at the right time in rich soil built up by the addition of manure and other organic matter.

Some people have reservations about applying animal manure and human wastes to the land, but the reply to this is very simple and clear. Life in nature is a continuous cycle between animals (man and livestock), plants, and microorganisms (the earth). Animals live by feeding on plants; the wastes excreted daily by these animals, and their bodies when they collapse and die, are buried in the soil where they become food for small animals and microorganisms in the soil—the process of rotting and decomposition. The microorganisms that abound in the soil live and die, supplying growing plants with nutrients that are absorbed through the plant roots. All three—animals, plants, and microbes—are one; they prey on each other and they also coexist and mutually benefit each other. This is the natural scheme of things, the proper order of nature.

Only man—a creature of nature—can be called a heretic. If he is to be regarded as unclean, then perhaps he should be removed outside of the natural order. But in all seriousness, man, as a mammal, and his wastes, as a part of normal nature, must be permitted to take part in the workings of nature. Primitive societies grew vegetables naturally next to their simple homes. Children played under fruit trees in the garden. Pigs came and poked at the stools left behind and rooted up the earth. A dog chased the pigs away and people scattered vegetable seed in the rich earth. The vegetables grew fresh and green, attracting insects. Fowl came and pecked at the insects, laying eggs that the children ate. This was still a common sight in farming villages throughout Japan until about a generation ago. Not only was this way of living the closest to nature, it was also the least wasteful and most sensible.

To view such extensive vegetable gardening as primitive and irrational is to miss the point. Lately, it has become popular to grow "clean" vegetables in greenhouses without soil. Plants are grown using gravel culture, sand culture, hydroponics, liquid nutrient culture, and by irrigating or spraying nutrient-containing water. People are making a big mistake if they intend in this way to grow "clean," microbe-less vegetables free of insect damage without using animal or human wastes.

Nothing is less scientific and complete than vegetables grown artificially using chemical nutrients and sunlight filtered through glass or vinyl panels. Only those vegetables grown with the help of insects, microbes, and animals are truly clean.

Scattering Seed on Unused Land: What I mean by the "semi-wild" cultivation of vegetables is a method of simply scattering vegetable seed in a field, orchard, on earthen levees, or on any open, unused land. For most vegetables, mixed sowing with ladino clover gradually gives a vegetable garden with a cover of clover. The idea is to pick a good time during the sowing season and either scatter or drill a seed mixture of clover and many vegetables among the weeds. This will yield surprisingly large vegetables.

The best time to sow vegetables in the autumn is when weeds such as crabgrass, green foxtail, wheatgrass, and cogon have matured and started to fade, but before the winter weeds have begun to germinate. Spring-sown vegetables should be seeded in late March and April after the winter weeds have passed their prime but before the germination of summer weeds. Winter weeds include paddy weeds such as water foxtail and annual bluegrass, and field weeds such as chickweed, bog stitchwort, speedwell, common vetch, and hairy vetch. When vegetable and clover seed are scattered among the still-green weeds, these act as a mulching material in which the sown seeds germinate with the first rain. However, if not enough rain falls, the germinated seedlings may be done in by sunny, dry weather the next day. One trick here then is to sow the seed during the rainy season. Leguminous plants are especially prone to failure and unless they grow quickly risk being devoured by birds and insects.

Most vegetable seeds germinate quite easily and the young seedlings grow more

vigorously than generally thought. If the seeds sprout before the weeds, the vegetables become established before the weeds and overwhelm them. Sowing a good
quantity of fall vegetables such as *daikon*, turnip, and other crucifers will hold
back the emergence of winter and spring weeds.

When left in the orchard until the following spring, however, these flower and
age, becoming something of a nuisance in gardening work. If a few of these vegetables are left to grow here and there, they will flower and drop seed. Come June
or July, the seeds will germinate, giving many first-generation hybrids close by the
original plants. These hybrids are semi-wild vegetables that, in addition to having a
taste and appearance quite different from that of the original vegetable, generally
grow to absurdly large proportions: great big *daikon*, turnips too large for children
to pull up, giant Chinese cabbages, crosses between black mustard and Indian
mustard, Chinese mustard and Indian mustard, . . . a garden of surprises. As
food, they are likely to overwhelm and many people may be hesitant about sampling, but depending on how they are prepared, these vegetables can make for very
flavorful and interesting eating.

In poor, shallow soil, growing *daikon* and turnips sometimes look as if they are
ready to roll down the hill, and the only carrots and burdock that can be grown
have a short, thick, sinewy root with many root hairs. But their strong, pungent
flavor makes these the very best of vegetables. Once planted, hardy vegetables such
as garlic, scallion, leek, honewort, dropwort, and shepherd's-purse take hold and
continue producing year after year.

Leguminous vegetables should be included in the seeds sown among the weeds
in spring to early summer. Of these, vegetables such as asparagus bean, cowpea,
and mung bean are especially good choices because they are inexpensive and high-
yielding. Birds will feed on the seeds for garden peas, soybeans, adzuki beans, and
kidney beans, so these must be encouraged to germinate very quickly. The best
way to get around this is to sow the seed in clay pellets.

Weak vegetables such as tomatoes and eggplants tend to become overwhelmed
at first by weeds. The safest way to grow these is to raise young plants from seed
and transplant them into a cover of clover and weeds. Rather than training tomatoes and eggplants into single-stem plants, after transplantation they should be left
alone and allowed to grow as bushes. If, instead of supporting the plant upright
with a pole, the stem is allowed to creep along the ground, this will drop roots
along its entire length from which many new stems will emerge and bear fruit.

As for potatoes, once these are planted in the orchard, they will grow each year
from the same spot, crawling vigorously along the ground to lengths of five feet
or more and never giving in to weeds. If just small potatoes are dug up for food
and some tubers always left behind, there will never be any want of seed potatoes.

Members of the Gourd family such as bottle gourd and chayote may be grown
on sloping land and allowed to climb up tree trunks. A single hill of overwintered
chayote will spread out over a 100-square-yard area and bear 600 fruits. Cucumber
should be of varieties that trail well along the ground. The same is true for melons,
squash, and watermelons. These latter have to be protected from weeds at the
seedling stage, but once they get a little larger, they are strong crops. If there is
nothing around for them to climb, scattering bamboo stalks with the tops remain-

ing or even firewood will give the vines something to grasp onto and climb; this benefits both plant growth and fruit production.

Yam and sweet potato grow well at the foot of the orchard shelterbelt. These are especially enjoyable because the vines climb trees and produce fairly large tubers. I am currently growing sweet potato vines over the winter to achieve large harvests. If I am successful, this will mean that sweet potatoes can be grown also in cold climates.

With vegetables such as spinach, carrot, and burdock, seed germination is often a problem. A simple and effective solution is to coat the seeds with a mixture of clay and wood ashes or to sow them enclosed in clay pellets.

Things to Watch Out for: The method of semi-wild vegetable cultivation I have just described is intended primarily for use in orchards, on earthen levees, and on fallow fields or unused land. One must be prepared for the possibility of failure if the goal is large yields per unit area. Growing one type of vegetable in a field is unnatural and invites disease and pest attack. When vegetables are companion-planted and made to grow together with weeds, damage becomes minimal and there is no need to spray pesticides.

Even where growth is poor, this can generally be improved by seeding clover together with the vegetables, and applying chicken droppings, manure, and well-rotted human waste. Areas unfavorable for vegetable growing are generally not conducive to weed growth, so a look at the types and amount of natural weed growth on the land can tell a lot about soil fertility and whether there are any major problems with the soil. Taking measures to bring about a natural solution to any problem may make it possible to produce a surprisingly rich growth of enormous vegetables. Semi-wild vegetables have a pungent aroma and good body. Because these have been produced in healthy soil containing all the necessary micronutrients, they are without question the most healthy and nutritious food man can eat.

By following the crop rotation system described earlier and growing the right crop at the right time, it may even be possible to grow vegetables in a semi-wild state over a large area.

Disease and Pest Resistance

Vegetable gardening in Japan has traditionally consisted of intensive cultivation in small garden plots for home consumption. The main sources of fertilizer were chicken droppings, livestock manure, human wastes, ashes from the furnace, and kitchen scraps. Pesticides were rarely used, if ever. In fact, pesticide use on the scale we see today is really a very recent phenomenon.

Recently, I came upon an old, dust-covered booklet I had written—and forgotten about—long ago while I was at the Kochi Prefecture Agricultural Testing Station during the war. It was entitled, "Proposal for the Control of Disease and Pest Damage in Vegetables."

I had written it as a practical manual for anyone intending to study disease and pest damage on their own. It contained reference tables on diseases and insect pests for different vegetables, and gave the most detailed possible descriptions of individual diseases and pests, the characteristics of pathogenic microbes, infection in plants, and the stages of development and behavior of insect pests. The methods of control I described in the booklet were all primitive and consisted almost exclusively of skillful trapping or some form of repulsion. There was almost nothing to write about insecticides. The agents most widely used at the time were herbs such as pyrethrum, tobacco, and derris root. Aside from this, lead arsenate was used in a very minor amount. Bordeaux mixture was used as a universal remedy for bacterial and fungal diseases, and sulfur preparations saw occasional use against certain diseases and mites.

Now that I think of it, it was fortunate that there were no pesticides at the time, for this allowed farmers and agricultural technicians to learn the characteristics of crop diseases and pests, and concentrate on preventing damage by these through repulsion and sound farming practices.

Today, with pesticides uniformly produced in massive quantities, growing vegetables without pesticides seems to many unthinkable, but I am convinced that by reviving the pest control measures of the not-so-distant past and practicing semi-wild cultivation, people can easily grow more than enough vegetables for their own consumption.

With the vast number of diseases and insect pests about, many farmers believe control to be impossible without proper expertise and pesticides. Yet, although from ten to twenty types of pests and diseases generally attack any one kind of vegetable, the only ones that are really major pests are cutworms, borers, leaf beetles, certain types of ladybugs, seed-corm maggots, and aphids. The others can generally be controlled by proper management.

Farmers a while back almost never used pesticides on vegetables in their kitchen gardens. All they did was to catch insects in the morning and evening on some gummy earth at the end of a piece of split bamboo. This worked well for caterpillars feeding on cabbage and other leaf vegetables, melon flies on the watermelon and cucumbers, and ladybugs on the eggplant and potatoes. Disease and pest damage to vegetables can usually be prevented by being familiar with the nature and features of such damage rather than attempts at control, and most problems can be taken care of by practicing a method of natural farming that gives some thought as to what a healthy vegetable is. Because hardy varieties are used, the right crop is grown at the right time in healthy soil, and plants of the same type are not grown together. Companion-planting vegetables of many different types in place of weeds in an orchard or on idle land is an eminently reasonable method of cultivation.

As an additional precaution, I would also recommend that pyrethrum and derris root be planted at the edge of the garden. Tests were conducted on different varieties of derris root at the Kochi Testing Station before the war, and those varieties that are cold-hardy, suited to outdoor cultivation, and have a high content of the active ingredient selected for use. Pyrethrum flowers and derris root may be dried and stored as powders. Pyrethrum is effective against aphids and caterpillars, while derris root works well against cabbage sawflies and leaf beetles.

However, these may be used against all insect pests, including melon flies, by dissolving the agent in water and sprinkling the solution onto the vegetable plants with a watering can. Both agents are harmless to man and garden vegetables.

While working in Kochi Prefecture, I remember seeing local chickens black as ravens strutting through a vegetable patch in a farmyard and deftly picking at insects without scratching the earth or harming the vegetables. Letting fowl loose in a vegetable patch can be one very effective way of keeping insect pests in check.

Try raising vegetables as the undergrowth in an orchard and let native fowl loose in the orchard. The birds will feed on the insects and their droppings will nourish the fruit trees. This is one perfect example of natural farming at work.

Resistances of Vegetables to Disease and Insects:

Strong (require no pesticides)

Yam family: Chinese yam, Japanese yam
Arum family: taro
Goosefoot family: spinach, chard, Chinese cabbage
Carrot family: carrot, honewort, celery, parsley
Composite family: burdock, butterbur, lettuce, garland chrysanthemum
Mint family: perilla, Japanese mint
Ginseng family: udo, ginseng, Japanese angelica tree
Ginger family: ginger, Japanese ginger
Morning-glory family: sweet potato
Lily family: Chinese leek, garlic, scallion, Nanking shallot, Welsh onion, onion, dogtooth violet, asparagus, lily, tulip

Moderate (require little pesticides)

Pea family: garden pea, broad bean, adzuki bean, soybean, peanut, kidney bean, asparagus bean, Egyptian kidney bean, sword bean
Mustard family: Chinese cabbage, cabbage, daikon, turnip, Indian mustard, rapeseed, leaf mustard, potherb mustard, sea-kale, black mustard

Low (require pesticides)

Gourd family: watermelon, cucumber, Oriental melon, pickling melon, squash, white gourd, chayote, bottle gourd
Potato family: tomato, eggplant, potato, red pepper, tobacco

Minimal Use of Pesticides: In principle, pesticides should not be used in natural farming, but at times there may be no alternative. The following chart is a simple guide for compounding pesticides and their proper and safe use.

Pesticide compounding chart

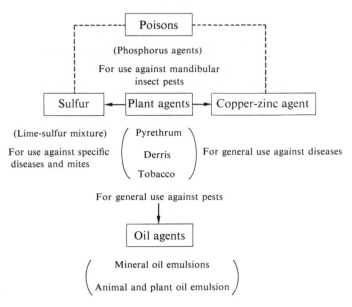

Poisons

(Phosphorus agents)

For use against mandibular
insect pests

Sulfur ← Plant agents → Copper-zinc agent

(Lime-sulfur mixture)

For use against specific
diseases and mites

Pyrethrum

Derris   For general use against diseases

Tobacco

For general use against pests

Oil agents

Mineral oil emulsions

Animal and plant oil emulsion

For use on sucking insect pests

← Mixes well

---- Mixes

Other mixtures not possible

The Road
Man Must Follow

1.   The Natural Order

Organisms of all manner and form inhabit the earth's surface. Broadly divided into animals, plants, and microorganisms, they differ from each other but are united in a single community of organic interrelationships. Man characterizes these interrelationships either as a competitive struggle for dominance and survival or as cooperation and mutual benefit. However, from an absolute perspective, these are neither competitive nor cooperative, but one and the same.

All living things belong to an endless food chain; all live by feeding on something and die at the hand of something else. This is the proper order of living nature. Matter and energy on the earth's surface are also in a constant state of flux, passing through continuous cycles without birth or death. Such is the true image of the universe.

Plants that grow on the earth are fed upon by bird and beast. Some of these animals become prey to other animals, while others eventually succumb to disease or age. Their wastes and remains are broken down by microorganisms which in turn proliferate and die, returning to the earth nutrients that are taken up once more by plants.

Among the microorganisms there are the bacteria, fungi (including the true fungi and molds), slime molds, and yeasts. Predator-prey relationships exist among members of this vast group as well. There are fungi that wrap mycelia about their prey and kill it by dissolution, bacteria which secrete substances that kill fungi, bacteriophages that kill bacteria, and viruses that kill both bacteria and fungi. Some viruses kill other viruses. And there are viruses, bacteria, and fungi that parasitize and kill plants and animals.

The struggle for survival among animals is identical. There are spiders that kill the rice borers and leafhoppers that feed on rice, mites that kill the spiders, predaceous mites that feed on these mites, ladybugs that feed on predaceous mites, earwigs that feed on the ladybugs, cricket moles and centipedes that eat the eggs of earwigs, swallows that feed on centipedes, snakes that eat small birds, and kites and dogs that kill snakes.

Bacteria and viruses attack these birds, beasts, and insects. Amoeba and nematodes feed on the bacteria, and the remains of nematodes are fed on in turn by earthworms, which are relished by moles. Weasels feed on the moles, and microorganisms break down the carcass of the weasel, providing a nutrient source for plants. The plants are parasitized by various pathogens, fungi, and pests, and serve as food for animals and man. The natural ecosystem is therefore an incredibly complex array of interdependently linked organisms, none of which live separate from the rest, none of which simply die and are done with. This must not be seen as a world of intense competition for survival or of the strong eating the weak, but as a united family of many members that live together in a single harmony.

232

Fig. 5.1 Cycles of the natural world.

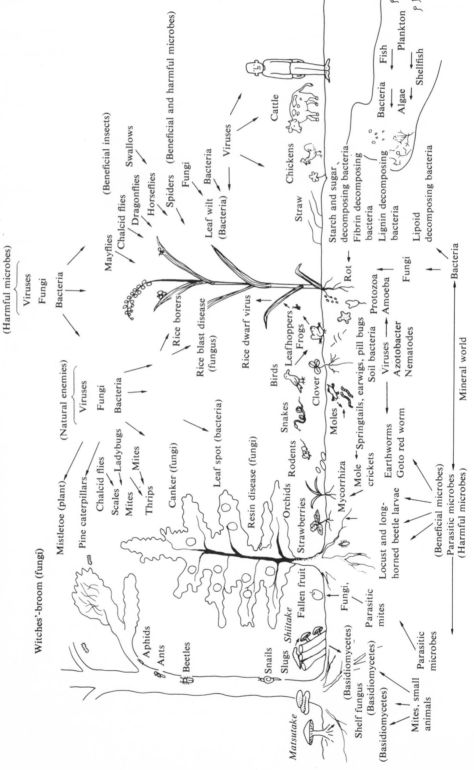

Microbes as Scavengers

The farmer dreads nothing more than to be caught loafing and despised by others for he will be told, "Don't think that you can live all by yourself. There are days of darkness too. When you die, you'll need the services of four people." However well we manage to get on without others in life, we always need four pallbearers at the funeral.

Actually, it takes more than four people to dispose of a corpse. Countless microbes and small animals in the soil are engaged in what could almost be called an assembly-line operation consisting of the dismantling, decomposition, rotting, and fermentation of the body. To completely return a corpse to the soil, billions upon billions of microorganisms appear one after another, making by turns the final service to a human being.

The days of man are filled with birth and death. A person's cells live on in his children and grandchildren where they continue multiplying day after day. At the same time, the body gradually breaks down, growing old and infirm. After death, the corpse is decomposed as food by bacteria, so one's ultimate form may be that of a microbial cell. And since the last to offer up incense to the departed soul are probably lactic acid bacteria, a person vanishes as a sweet, acrid aroma through lactic fermentation.

Thanks to the microbes that dispose of the remains of animals and plants, the earth's surface is always clean and beautiful. If animals died and the carcasses just remained there without decomposing, a couple of days is all it would take to make the world an intolerable place to be. People look on casually at this activity by microbes and small animals, but there is no greater drama in our entire world.

No species of bird flying in the air should become extinct. The earthworms that burrow through the soil must not vanish. Nor must mice and spiders proliferate too greatly. If one type of fungus thrives just a little too well, this throws everything out of balance. Tens of millions of species live on in perfect harmony without increase or decrease; they are born and they die unseen by man. The mastery of the conductor performing this drama of natural transformation at once casual and intense is truly something to behold. What can the mechanism be by which all the organisms of the world propagate in just measure—becoming neither too numerous nor too scarce? Such natural, self-governed providence is indeed a mystery.

But there is one who disrupts this natural order. It is man. Man is the sole heretic in the natural order. Only he acts as he pleases. Instead of burying his corpses in the earth, he douses them with heavy fuel oil and cremates them. Much is made of the sulfurous gases discharged from the crematorium chimney, but the polluting effects on other smaller animals and plants are surely greater than the effects on man. People think of cremation as fast, convenient, and hygienic because a corpse can be entirely disposed of in just two hours. But what about the fuel oil that is mined, transported, and burned in the crematorium furnace? If this and anti-pollution treatment of the stack gases are taken into account, cremation is neither fast nor clean. Perhaps simple burial or open burial in which the corpse is left exposed to the elements may seem primitive and inefficient to the short-sighted, but these are the most sensible and complete ways to dispose of a corpse.

Designs for the most advanced refuse processing plants are child's play compared with the infinitely elaborate methods of treating garbage used by nature. Human society almost has more than it can handle in just treating the garbage that issues from the kitchen, but nature works on a totally different scale.

It takes from twenty minutes to an hour for one bacterium or yeast to divide and become two, and the same amount of time for two to become four. Assuming multiplication to continue unchecked in the presence of food and suitable temperatures and humidity, after two or three days, a single bacterium such as *Escherichia coli* could leave a mass of progeny equivalent to the total mass of all living organisms on the face of the earth. This means that if the self-governing mechanisms by which nature regulates and controls the reproductive power of one type of bacteria were to cease operating for just several days, the earth would become a morass of bacterial remains. The ability of the earth's organisms to multiply is far more powerful than people imagine. At the same time, their ability to destroy and dispose of organisms is also very great.

The balance between multiplication and destruction, the equilibrium between production and consumption, the fact that nature has seen to the growth and propagation of organisms and also to the treatment of their wastes and remains, carrying out both rapidly and harmoniously without the least hitch for untold thousands and millions of years, all this is of enormous consequence. It is here that one must turn for a just comparison of the powers of man with those of nature.

A look at how nature disposes of the carcass of an animal will show a method that is perfect in every sense—biologically, physically, and chemically. If man were to try the same thing himself, his method would be plagued with problems and would invariably create pollution in some form.

I would like to give one more example of just how amazing nature is when we take even a casual look at what transpires there. I remember searching once, while at the Kochi Prefecture Agricultural Testing Center, for a beneficial bacterium with which to prepare compost from straw and brush. I needed a bacterium capable of quickly decomposing straw and other coarse plant material. This was something like the beneficial bacteria scientists search for today to convert garbage and sludge into artificial manure for use as fertilizer.

I collected refuse from garbage sinkholes as well as cattle, hog, chicken, rabbit, and sheep droppings. From these, I isolated and cultured microorganisms, obtaining samples of many different bacteria, fungi, slime molds, and yeast. I was able in this way to collect samples of many microbes suited to preparing compost. I then inoculated samples of each of these into straw in test tubes or within concrete enclosures and observed the rotting rates.

Later, however, I realized that such an experiment was really quite worthless. To one concerned with how long things take, an investigation such as this may seem useful, but a closer look reveals that nature makes use of far better methods of treating garbage and preparing compost.

Rather than going to all the trouble of isolating beneficial microbes and inoculating straw with this "fermentation promoter," all I had to do was scatter a handful of chicken droppings or clumps of soil over the straw. Not only was this the quickest way, it also gave the most complete rotted compost.

There is no call for making a lot of fuss over "microbial" and "enzymatic" farming methods. The following transformations take place on a rice straw casually dropped onto the earth.

The straw draws a lot of flies and other small insects that lay eggs from which maggots and other larva soon emerge. Before this happens, however, rice blast disease, leaf blotch, and rot-causing fungi already present on the rice leaves spread rapidly over the straw, but spider mites are soon crawling over this fungal growth. Next, different microbes begin to proliferate at once. The most common include yeasts, blue mold, bread mold, and trichoderma fungi, which destroy the pathogens and begin to decompose the straw. At this point, the number and types of organisms drawn to the straw increase. These include nematodes that feed on the fungi, bacteria that feed on the nematodes, mites that consume the bacteria, predaceous mites that feed on the mites, and also spiders, ground beetles, earwigs, mole crickets, and slugs. These and other organisms mingle and live in the straw, which undergoes a succession of "tenants" as it gradually decomposes.

Once the fibrin-decomposing fungi run out of food, they stop growing and are supplanted by lipoid- and lignin-decomposing bacteria which feed on the fungi and the scraps left over by the fungi. Before long, parasitism and cannibalism sets in among the aerobic bacteria, and these are gradually replaced by anaerobic bacteria. Lactic acid bacteria round off the process with lactic acid fermentation, at which point all trace of the straw disappears. This is just the briefest of looks at the total decomposition of a single piece of straw on the ground over a period of several days.

Microbiologists are well aware of how rapidly and perfectly the processes of decomposition and rotting break down garbage in the natural world. Yet man, believing that he has to make intensive use of beneficial microbes to speed up putrefaction or that he must raise the temperature to promote bacterial growth, prepares compost. He should stop and consider how worthless and undesirable such efforts are. Frankly, anything that he does just disturbs the rapid and perfect natural processes.

People must not forget, in looking at the rotting of a straw, at the fertilizer response, at soil improvement, and at all the other processes that take place in nature, that what man knows is only the most minute, infinitesimal part of the natural order. In addition to the very visible lead roles are an infinite number of supporting roles that perform important yet unknown functions. If man jumps onto center stage and begins giving out directions like a know-nothing director, the play will be ruined. When something goes wrong in nature, the biosystem changes course. Unlike in a factory where the damage may consist of only a broken gear, in nature a disruption gives rise to an unending series of repercussions.

Plants and animals seem to live freely and without constraints of any sort, but in fact they belong to a close-knit order. Man casts stones into this order, the biggest of which are pesticides, fertilizers, and agricultural machinery. He goes ahead and uses pesticides, for example, because of their ability to destroy specific pests or pathogens, but is almost totally ignorant of the ripple effects pesticides have on the rest of the biological world.

Here, as a small example, is an incident that occurred locally. My village is noted for its Karakawa loquats. Once, as I was touring the village with officials from the local agricultural cooperative, we passed in front of a loquat orchard and I remember someone saying, "This year's loquats have been hit again by the cold and aren't blooming well at all. With this happening year after year, farmers are losing interest in growing loquats." Finding this a bit hard to believe, I stopped the car and went in to check the orchard. I found most of the flower corollas rotted and noticed on these the spores of a botrytis fungus. Explaining that this was not cold-weather damage but a botrytis disease, I described how the problem could be taken care of by spraying and suggested two or three ways of doing so. The astonished head of the horticulture cooperative immediately got in touch with the local agricultural testing station and, with the whole village cooperating in a pesticide spraying program, the organism was soon brought under control.

The loquats gradually came back and everything appeared to look rosy again, but one question remained unanswered. Why had this outbreak occurred in the first place? My theory is that it was triggered by the sudden spraying of a whole series of new pesticides following the war in an attempt to control citrus diseases. This may appear far-fetched to some, but here is how I came to my conclusions. I cannot be absolutely certain as I did not run any laboratory experiments on this, but I believe the organism responsible to have been a botrytis fungus of as-yet undetermined identity. Either it belonged to the species *Botrytis cinerea*, which causes gray mold in citrus fruit, or was a variant of the same. Based on this supposition, the severe outbreak of gray mold may have been caused by the following:

1) Interplanting of citrus trees in loquat groves due to the boom in citrus production.
2) The rapid transition in local orchards from clear cultivation to mulching and sod cultivation, creating a soil surface environment of increased moisture ideal for the propagation of microbes.
3) Promotion of the practice of thinning fruit. Young fruit were dropped to the ground, and there colonized by the fungi.
4) Use of the Bordeaux mixture, which is effective against fungi, was discontinued and new pesticides used in its place.

This fungus is partially saprophytic and inflicts serious damage when present in large numbers. Causes for emergence are usually poor orchard sanitation, excess humidity, low tree vigor, and entanglement of the branches and foliage. Since, of these, the single largest factor is the microclimate in the orchard, the chief cause of the fungus outbreak was probably excess humidity. If this is the case, then I was partly to blame.

Immediately after the end of the war, I encouraged farmers, as part of a public campaign for eliminating widespread malnutrition, to sow clover in the citrus orchards and idle village land and to raise goats. This practice caught on quite well and in many cases resulted in sod orchards. The high humidity in these orchards may very well have been a cause for the proliferation of gray mold and rotting of the loquat blooms. If so, the farmers had sown the seeds of their own misfortune, but the one most responsible may have been me.

The matter does not end here. Having identified the problem as a botrytis disease and sprayed with strong pesticides such as zineb, organoarsenic, or organochlorine agent and applied herbicides, farmers are now rejoicing that the disease has been brought under control, but do they really have cause for celebration?

The fungus remains dormant throughout the winter in the corollas of fallen flowers, following which the hyphae fuse to form a sclerotium about the size of a poppy seed. A small mushroom forms within this sclerotium, and in the mushroom is formed an acospore, or spore-containing sac. This sac, which measures less than one millimeter across, contains eight tiny, genetically distinct spores. If the acospores of this fungus are octopolar, then it may be capable of producing more variants than even the tetrapolar *shiitake* fungus.

What I mean to show by all this is that, although new strains of advanced animals and plants do not arise easily, the chances of this happening in lower bacteria and fungi are very great and can lead to frightening consequences. Spraying pesticides with high residual toxicity and mutagenic chemicals onto easily mutated microbes is asking for trouble, for who knows what strange mutants may arise.

The result may very well be new pesticide-resistant pathogens and highly parasitic microbes. Another personal experience showed me just how possible this is. Because the resin disease fungus that attacks lemon and grapefruit trees grown in the United States and the fungus that attacks Satsuma oranges and summer oranges in Japan bear different scientific names, I thought they were different species, but when I tried crossing them, mycelial conjugation took place and acospores were formed. By crossing these eight spores in various ways, I was able to produce different strains.

Leave Nature Alone

People might object to new strains of pathogens, but to the scientist these are of great fascination. Conversely, there is no way of telling when something that is beneficial to man today may suddenly become harmful. Apart from the basic stance of not opposing nature, there are no absolute criteria for judging what is good or bad, what is an advantage and what a liability. Although the common rule is to make such judgments on a case-by-case basis under the imperatives of the moment, nothing could be more dangerous.

As the use of new pesticides grew more widespread following the war, reports of major outbreaks of pesticide-resistant pathogens and pests suddenly started appearing. Dozens of organisms were involved, including mites, leafhoppers, rice borers, and beetles. Although one possible explanation is the selection and survival of organisms resistant to the pesticides, another possibility is that hardy organisms adapted to pesticides arose. Even more frightening is the distinct possibility that the use of pesticides may have triggered the emergence of ecospecies and mutants. Some scientists are concerned about the chances of a "retaliation" by insects, but I believe that much more is to be feared from bacteria, fungi, and viruses.

New pesticides about which only the degree of toxicity in the human body is investigated, breeding experiments for the creation of new plant varieties through radiation . . . . Scientists believe they are wrestling in earnest with the problem of pollution when in fact they are just sowing the seeds for future pollution.

When the various plants in a field are doused with radiation, the scientists running such experiments give no thought to the changes this effects in the soil and airborne microbes. As I watched a show about these experiments on television not too long ago, I was far more concerned with the microbial mutants and spores that could reasonably be expected to arise in such an irradiated field than with admiration or expectations over what new and aberrant types of plants might result. Being invisible to the unaided eye, it is harder to tell whether any monstrous microbes have been created.

Monsters belong to the world of comics, but don't they already exist in the microbial world? With the development of rockets and space shuttles, no scientist would guarantee that there is no danger of non-terrestrial microbes being brought to earth from the moon or other heavenly bodies. What is unknown is unknown. If an organism exists that cannot be detected by terrestrial methods of identification, then there is no way to quarantine it. Verification that an organism originated from a heavenly body is not likely to occur until it has fluorished on the earth. How can man hope to correct the accidents in the biosphere that have begun happening about us and the abnormalities we are seeing in the natural cycles?

Although I have no way of knowing for certain, I suspect that what happened was that atmospheric pollution killed off microbes that attack various botrytis fungi, and that this triggered the rotting of apple, loquat, and plum blossoms, and a massive outbreak of gray mold on citrus fruit. The explosive increase in this mold led to a sudden rise in nematodes that feed on the mold, resulting in an abnormally large increase in the number of dead pine trees. This prolific gray mold was also responsible for the destruction of the *matsutake* fungi that live parasitically at the roots of the pine trees.

The true cause may be unclear, but one thing is certain: an inauspicious change has overtaken the strongest form of life on the Japanese archipelago—the Japanese red pine, and the weakest form of life—the *matsutake* fungus.

2.　Natural Farming and a Natural Diet

Agriculture arose from human cravings over food. It was man's desire for tasty and abundant food that was responsible for the development of agriculture.

Farming methods have constantly had to adapt to changes in the human diet. Unless the diet is basically sound, agriculture too cannot be normal.

The Japanese diet has undergone rapid development recently, but has this really been for the good? The failure of modern agriculture has its roots in abnormal dietary practices and the low level of basic awareness people have regarding diet.

What is Diet?

The very first step that must be taken in setting the proper course for agriculture is to reexamine what "diet" represents. Correcting man's eating habits by establishing a natural diet erects a foundation for natural farming.

Has man been correct in developing systems of agriculture based on his eating habits or was this a major error? Let us turn back to examine the driving forces behind the development of diet: the original cravings of man, the sense of starvation, the emotions which cry out that food is scarce, the will that seeks after plenty.

Primitive man fed himself on whatever he could find about him—vegetation, fish and shellfish, fowl and animals; everything served a purpose, nothing was useless. Most things served either as food or as medicine. There was surely more than enough food to feed the entire human population of the earth.

The earth produced in abundance and food enough to satisfy everyone was always to be had. Had this been otherwise, man would not have emerged on the face of the earth. The smallest insects and birds are provided with more than enough food without having to cultivate and grow some for themselves. How odd then that only man laments over a want of food and frets over an imbalance in his diet. Why, under circumstances where the lowest of organisms thrived quite well, did only man become concerned over diet and feel compelled to develop and improve food production?

Animals are born with an instinctive ability to distinguish between what they can and cannot eat, and so are able to partake fully of nature's plentiful stores. In man, however, the stage of infancy during which he feeds instinctively is short. Once he starts to become familiar with his surroundings, he makes judgments and feeds selectively according to impulse and fancy. Man is an animal that feeds with his head rather than his mouth.

Scientifically, we characterize foods as sweet, sour, bitter, hot, tasty, unsavory, nutritious, unnourishing. But what is sweet is not always sweet, nor is something tasty always tasty. Man's sense of taste and his values change constantly with time and with the circumstances.

When we are full, the most delectable food is unpalatable, and when we are hungry, the most awful-tasting food is delicious. Nothing tastes good to a sick man and nothing is nutritious to one who is not healthy. Unconcerned over whether taste is associated with the food proper or the person eating it, man has elected to produce food with his own hand. Differentiating among foods and calling them sweet or sour, bitter or hot, tasty or bad-tasting, he has gone in pursuit of flavors that please the palate, letting his fancy take the better of him. This has resulted in an unbalanced and deficient diet. Also, as he has selected the foods that suit his taste, man has lost the native intelligence to partake of what is really necessary to him.

Once man eats something sweet, food that he had felt until then to be sweet loses its appeal. Once he samples epicurean food, plainer fare becomes unacceptable and he goes off in search of even greater culinary extravagance. Unconcerned about whether this is good or bad for the body, he eats according to the dictates of his palate.

The food that animals eat by instinct constitute a complete diet, but man, with his reliance on discriminating knowledge, has lost sight of what a complete diet is. As the harm caused by an unbalanced diet becomes clear, man grows concerned over the incompleteness and contradictions in his diet. He attempts to resolve this through science, but the desires from which spring his cravings proceed one step ahead of these efforts, aggravating the problem.

As man works to correct his unbalanced diet, he studies and analyzes food, calling this a nutrient, that a calorie, and trying to combine everything into a complete diet. This seems to bring him closer to his goal, but the only real outcome of his efforts is the fragmentation of diet and even greater contradiction. Someone that has no idea of what a complete diet is cannot rectify an unbalanced diet. His efforts never amount to more than a temporary solace. The best solution would be to find a complete diet that satisfies human cravings, but this will never happen.

Scientific investigations on food are confined to analytic research. Food is broken down into a limitless array of components—starch, fat, protein, vitamins A, B, C, D, E, F, B_1, B_2 and so on, and each studied intensively by specialists. But this process has no conclusion, resulting only in infinite fragmentation.

We can safely say that what primitive man ate instinctively comprised a complete diet. On the other hand, instead of leading us toward a complete diet, modern science has resulted in the discovery of a more sophisticated but imperfect diet. Man's quest for a complete diet has led him in the opposite direction.

Although the development of new foods that satisfy human cravings continues, such cravings are merely illusions spun by man over things in the phenomenological world. These illusions invite other illusions, widening the circle of human delusion. The day that these cravings are fully satisfied will never come. Indeed, the rapid advance of his cravings and desires only increases man's frustrations. No longer content with food available close at hand, he travels off in search of whales in the south seas, marine animals in the north, rare birds in the west, and sweet fruit in the east. Man goes to no ends to satisfy the cravings of his palate.

Although he could have lived quite well just by working a tiny strip of land, he now rushes about in a frenzy because there is no food, or the food is bad-tasting, or delicious, or unusual. What it amounts to is that the entire world is rushing about to lay their hands on choice foods.

If these were really delicious, then one could understand all the activity; if favorites such as liquor, cigarettes, and coffee were really as good as they are made out to be, then nothing could be done about it. But the fact remains that, no matter how enjoyable these may be, they have never been essential to the human body. Tastiness exists in the minds of people who believe something to taste good. The absence of delicacies does not prevent the feeling of "deliciousness" from arising. People who do not consume delicacies may not experience ecstasy at the dinner table as often, but this does not mean they are unhappy. Quite the contrary.

A look at the food industry, which has worked tirelessly to develop new popular foods and a complete diet, should give a clear idea of the likely outcome of the progress that man strives for. Just look at all the food products flooding the stores. Not only are there full assortments of vegetables, fruits, and meat in all seasons, the shelves are overflowing with an endless variety of canned foods, bottled foods, frozen and dried foods, instant dinners packed in polyester bags. Is this vast array of food products, from raw foods to processed foods in a variety of forms—solid, liquid, powdered—with their complement of additives for tickling the palate, really essential to man? Does it really improve his diet?

This "instant" food that panders to consumer cravings and was created for greater rationality and convenience in the diet has already deviated far from its original goal. Food today is thought of less as something that supports life than as something to please the human palate and titillate the senses. Because it is "convenient" and "quick and easy to prepare," it is highly valued and produced in large quantities.

Man thinks he has made time and space his, but people today no longer have any time. This is why they are delighted with instant foods. As a result, food has lost its essence as real food and become only a concoction.

Yet, even so, some people believe that with further advances in food technology it will eventually be possible to produce complete instant foods in factories, liberating man from his tiresome dietary habits. Some even expect to see the day when one small food tablet a day will fill the stomach and sustain physical health. What utter nonsense.

A complete food for man that includes all the necessary nutrients in sufficient quantity must, in addition to containing every one of the components in the roots, leaves, and fruits of vegetation growing on the earth, in the flesh of all birds and beasts, fish and shellfish, and in all grains, must also have added to it some as yet unknown ingredients. Creating such a complete food would require incredibly huge expenditures of capital for research and production, not to mention long hours and great labor in sophisticated plants. The end product would be horrendously expensive, and far from being as compact as a pill, would probably be extremely bulky.

Those forced to eat such food would probably groan, "Complete food takes so much labor and time to produce. How much easier, cheaper, and tastier it used to be to eat raw food grown in the garden under the sun. I'd rather die than have to go on packing my guts with such strange, foul-smelling food as this."

People talk of eating delicious rice and growing delicious fruit, but there never really was anything like delicious rice in this world to begin with, and growing delicious fruit just adds up to a lot of wasted toil.

Tasty Rice

More than thirty years have passed since the days of famine and hunger in Japan following the end of World War II. Today, those times appear as but a dark dream of the past. With the bumper crops of grain we have been seeing over the last dozen years or so, rice surpluses have formed and there is no longer enough warehouse space to store all the old grain. Dissatisfied consumers are furious, complaining that the price of rice is too high, that they have no need for "bad-tasting" rice, that they want to eat "good-tasting" rice, that new and more palatable varieties of rice had better be produced. Politicians, traders, and the agricultural cooperatives representing farmers have added their voices to the angry din, pounding the desks and huddling together to come up with a hundred brilliant ideas. Agricultural technicians have been ordered to keep farmers from establishing new paddy fields and to encourage farmers to stop growing "bad-tasting" rice and grow "tasty" varieties instead or switch to other crops.

But this sort of controversy is possible only when people have no idea of the true nature of the food problem. This debate over "good-tasting" rice alone gives a clear view of the fantasy world of man. It might be helpful to consider whether tasty rice really exists in this world at all, whether the angry movement to secure such rice can really bring joy and happiness to man, and whether such a movement is worthwhile to begin with.

I do not mean to deny that "tasty" rice and "untasty" rice do not exist, only to point out that the difference in taste between different varieties is very small. For example, even were a farmer to select a good-tasting variety of rice and, sacrificing yields, willingly devote himself wholeheartedly to perfecting techniques for growing good-tasting rice, just how delicious would the rice that he grew be? No rice would win unanimous praise by a panel of samplers. And even if it did, the difference with other varieties would be very, very minor.

Tasty rice cannot always be produced from a tasty variety. It is far too simplistic to think that the original difference in taste between varieties will be sustained right up to the dinner table. Depending on the land on which it is grown, the method of cultivation, and the weather, poor-tasting varieties may approach tasty varieties in flavor, while tasty rice, when hit by bad weather and heavily attacked by disease and pests, is often less palatable than poor-tasting rice. The minor differences in taste between varieties are always subject to reversal. And even when it appears as if tasty rice has been produced, the taste may deteriorate during harvesting, threshing, or processing. The chances of a rice being produced that retains the inherent properties of that variety are less than one in several hundred.

As hard as the farmer may try to produce tasty rice, this taste can be destroyed or retained depending on how the rice dealer processes the grain. The dealer grades the rice from various farming districts, processes the rice by milling it to various degrees, and mixes it in given proportions to create hundreds of varieties with distinctive flavors. Tasty rice can be converted into tasteless rice, and tasteless rice into tasty rice. Then again, when the rice is cooked at home, whether one soaks the rice overnight in water and drains it in a bamboo sieve, how much water one uses, how high the flame, the type of fuel, and even the quality of the rice cooker can all have an effect on the taste of the rice. The difference between good- and bad-tasting varieties of rice and between old and new rice can fall either way depending on how the grain is processed and cooked. One could say that it is the farmer, the rice dealer, and the housewife who create tasty rice, but in a sense no one creates tasty rice.

Fig. 5.2 shows that, even if we consider just a few of the production conditions, the chances that a tasty variety of rice will be grown, properly processed, and cooked skillfully to give rice of outstanding taste is not more than one in a thousand. This means that, even with the best of luck, someone may encounter truly tasty rice perhaps once in every two or three years. And if that person does not happen to be very hungry at the time, all will have been for naught.

This campaign for tasty rice has placed a great burden on the farmer and forced the housewife to buy high-priced rice without knowing what is going on. The only one likely to benefit from all this is the merchant. Bitten by the illusion of slightly tasty rice, people today are all floundering about in a sea of mud and toil.

Fig. 5.2   Good-tasting rice is a figment of the imagination.

My thinking on natural diet parallels that on natural farming. Natural farming consists of adapting to true nature, that is, nature understood with non-discriminating knowledge. In the same way, a true natural diet is a way of eating where one feeds randomly with an undiscriminating attitude on food taken from the wild, crops grown by natural farming, and fish and shellfish caught using natural methods of fishing.

One must then abandon an artificial diet designed on the basis of discriminating scientific knowledge and, gradually liberating oneself from philosophical constraints, ultimately deny and transcend these.

However, knowledge useful for living may be permitted if it can reasonably be thought to have arisen from undiscriminating knowledge. The use of fire and salt may have been man's first step away from nature, but these were first used in cooking when primitive man perceived the wisdom of nature and were heavenly inspired.

Agricultural crops which for many thousands of years have merely adapted to the environment and at some point survived through natural selection to become fixtures of human society may be thought of as foods that arose naturally rather than as artificial foods which originated through the application of discriminating knowledge by the farmer. This of course does not apply to crops that have been developed more recently through genetic improvement and are considerably alienated from nature. These, along with artificially bred fish and livestock, should be firmly excluded from the diet.

A natural diet and natural farming are not separate and distinct ideas, but united intimately as one whole. They are one too with natural fishing and animal husbandry. Man's food, clothing, and shelter, and his spiritual existence must all be blended together with nature in perfect harmony.

Plants and Animals Live in Accordance with the Seasons: I drew Fig. 5.3 thinking that this might help one to understand a natural diet that encompasses the theories of Western nutritional science and the Eastern philosophy of yin and yang, but transcends both.

Here I have crudely arranged foods according to the colors of the four alternating seasons, based on George Ohsawa's principle of yin and yang. Summer is hot and yang, winter is cold and yin. In terms of light, summer is said to be represented by red and orange, spring by brown and yellow, fall by green and blue, and winter by indigo and purple.

The diet is such that a balance is maintained between yin and yang and the color arrangement is harmonic. Thus, in the summer (yang) one should eat yin foods, and in the winter (yin) one should eat yang foods.

Foods are represented by different colors: vegetables are green, seaweed is blue, cereal grains are yellow, and meat is red.

Meat is yang and vegetables yin, with grains in between. Because man is an omnivorous animal that is yang, this leads to a set of principles which says that, when grains, which are intermediate, are eaten as the staple, yin vegetables should be consumed and meat (very yang)—consumption of which is essentially cannibalism—should be avoided.

However, even if these principles are essential medically or in the treatment of disease, too much concern and attention over whether something is yin or yang, acidic or alkaline, and whether it contains sodium and magnesium and vitamins and minerals leads one right back into the realm of science and discriminating knowledge.

Fig. 5.3 Harmony in the natural diet.

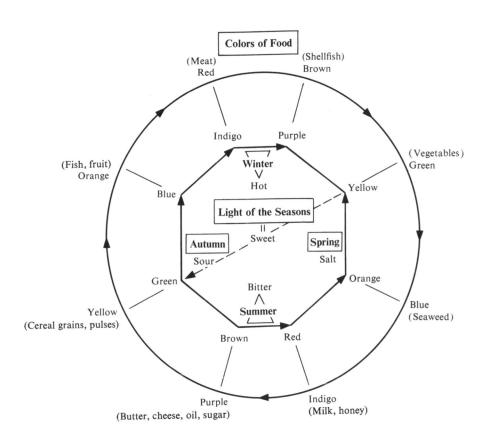

The mandala in Fig. 5.4 gives a somewhat systematic arrangement of foods readily available for consumption by man. This will give an idea of just how vast a variety of foods exists on the earth for man's survival. Those who live at the perimeter of spiritual enlightenment have no need to differentiate between any of the plants or animals in this world; all may become the exquisite and delectable fare of the world of rapture. Unfortunately, however, having alienated himself from nature, only man cannot partake directly of its bounty. Only those who have succeeded in fully renouncing the self are able to receive the full blessings of nature.

Fig. 5.4 Nature's food mandala—plants and animals.

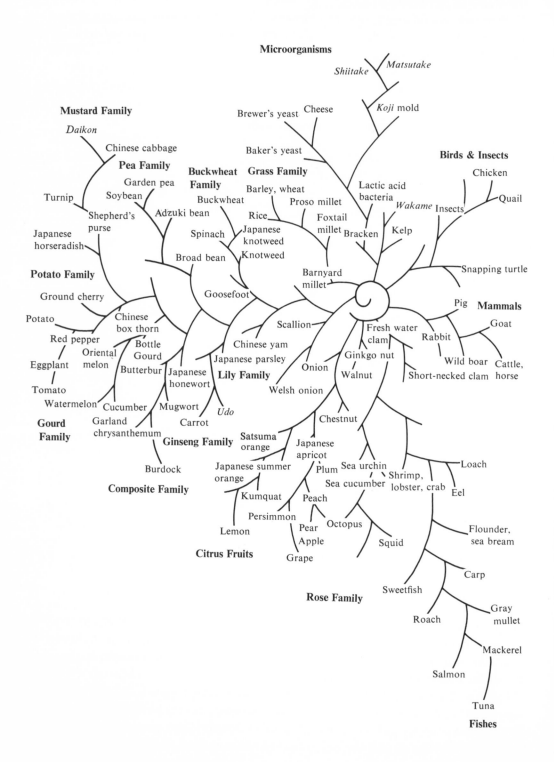

Fig. 5.5 is a mandala showing the foods available during each of the months of the year. This shows that, as long as man accepts and lives in accordance with the divine dispensation, a complete natural diet will arise of itself without his needing to know anything and without his having to ponder the principle of yin and yang. Of course, the foods consumed will vary with time and circumstances, and with the degree of health or malady.

Eating with the Seasons: The food that farmers and fishermen have taken locally for thousands of years is a splendid example of natural diet in accordance with the laws of nature. The seven herbs of spring—Japanese parsley, shepherd's purse, cudweed, chickweed, bee nettle, wild turnip, and wild radish—emerge early in the new year from the slumbering brown earth. As he enjoys the flavors of these herbs, the farmer meditates joyfully on his having survived a harsh winter. To go along with the seven herbs, nature provides shellfish—a brown food. The savory taste of pond snails, fresh water clams, and sea clams in early spring is a special treat.

A little later, in addition to such popular edible wild herbs as bracken and osmund, almost anything can be eaten, including young cherry, persimmon, peach, and Chinese yam leaves. Depending on how they are cooked, these may also serve as seasonings. Just as the first broad beans are ready for picking, edibles from the fields suddenly increase. Bamboo shoots are delicious with rockfish. Red sea bream and grunt can be caught in quantity and are excellent at the time of the barley harvest in late spring. Spanish mackerel *sashimi* in the spring is so good you want to lick your plate clean. During the festival of the Japanese iris, an offering is made of hairtail prepared with Japanese iris.

Spring is also a season for taking walks along the seashore, where seaweed—a blue food—is to be had. Loquats glistening in the early summer rains not only are a beautiful sight to see, this is a fruit that the body craves. There is a reason for this. All fruits ripen at the right time of the year, and this is when they are most delicious.

The time for pickling the green Japanese apricot (*ume*) is also one for enjoying the bracing flavor of the pickled scallion. This is when the rainy season lets up and summer makes its arrival. One quite naturally hankers for the fresh beauty and taste of the peach, and the bitter and sour flavors of the oleaster berry, plum, and apricot. Those who would refrain from eating the fruit of the loquat or peach have forgotten the principle of using the whole plant. Not only can the flesh of the loquat be eaten, the large seeds can be ground and used as coffee and the leaves can be infused to give a tea that serves as the best of all medicines. The leaves of the peach and persimmon give a potion for longevity.

Under the hot midsummer sun, one may even eat melon, drink milk, and lick honey in the cool shade of a tree. Rapeseed oil and sesame oil revive the body worn down by the summer swelter.

Many fruits ripen in early autumn, a time when yellow foods such as cereal grains, soybeans, and adzuki beans also become available. Millet dumplings enjoyed under the moonlight; taros and green soybeans cooked in the pod; corn-on-the-cob, red beans and rice, *matsutake* mushrooms and rice, and chestnuts and

248

rice in late autumn also make sense. And most welcome of all are the ripened grains of rice that have fully absorbed the yang of summer, providing a food staple rich in calories in preparation for the winter.

Barley, another staple that is slightly more yin than rice, is harvested in the spring and can be eaten as iced or hot noodles; it is almost uncanny how this suits the palate just as the appetite lags under the summer heat. The buckwheat harvested in late summer and early fall is a strongly yang grain, but is most essential during the summer.

Fig. 5.5 Nature's food mandala—the seasons.

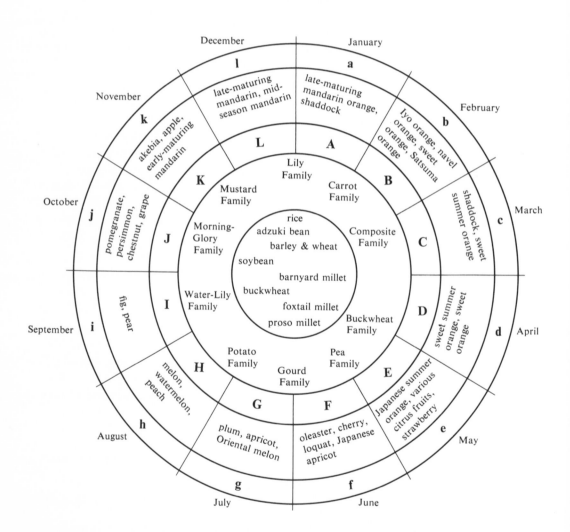

A. shallot, garland chrysanthemum, butterbur flower, creeping saxifrage, beet, lettuce, Indian mustard, Chinese cabbage, spinach, small turnips, burdock
B. Japanese parsley, honewort, celery, butterbur flower, *daikon*, Chinese cabbage, potherb mustard, Welsh onion
C. wild rocambole, leek, field horsetail, mugwort, spring *daikon*, scallion, comfrey, chard, lettuce, Indian mustard, carrot, seven herbs of spring, shallot
D. *shiitake*, leaf buds, Japanese pepper, Japanese angelica tree buds, *udo*, Chinese box thorn, osmund, bracken, Japanese knotweed, thistle, violet, Chinese milk vetch, aster, bamboo shoot, spring onion, Chinese cabbage, garland chrysanthemum, parsley, garden pea
E. wild rocambole, leek, perilla shoots, chard, cabbage, pepper, garden pea, broad bean, kidney bean, young turnip, bamboo shoot, butterbur, agar-agar, *wakame*
F. scallion, bracken, burdock (leaves), ginger (leaves), honewort, perilla (leaves), garden pea, asparagus, garlic, early-maturing green soybean, onion, young potato, summer *daikon*, spring-sown carrot, eggplant, cucumber
G. young turnip, okra, leek, Japanese ginger (flower), perilla (panicle), squash, eggplant, cucumber, summer *daikon*, Egyptian kidney bean, mid-season green soybean, onion, spinach
H. cucumber, squash, ginger, Chinese box thorn, knotweed, perilla (seed), winter melon, pickling melon, tomato, spring-sown burdock, cabbage, parsley, kidney bean, asparagus bean, early sweet potato
I. corn, arrowhead, autumn eggplant, green soybean, early-maturing taro, ginger, red pepper, *hatsutake*, *shimeji*, knotweed, sesame
J. mushrooms, *matsutake*, *shiitake*, lily bulb, shallot, honewort, garland chrysanthemum, sweet potato, soybean, peanut, taro, Chinese yam, lotus root, burdock, Welsh onion, Chinese cabbage
K. ginkgo nut, celery, chrysanthemum, green laver, *wakame*, *hijiki*, kelp, autumn *daikon*, summer-sown burdock, autumn potato, Indian mustard, Chinese cabbage, spinach, potherb mustard
L. Chinese yam, Chinese cabbage, leaf mustard, *daikon*, turnip, cabbage, summer-sown burdock, summer-sown carrot, onion, lotus root, arrowhead
a. edible fowl, snapping turtle, edible frog, oyster, sea urchin, sea cucumber, gray mullet, carp, river fish, sea bream, flying fish, herring
b. pond snail, sea cucumber, squid, mackerel, sardine, bluefish, Spanish mackerel, yellowtail
c. short-necked clam, clam, fresh water clam, river trout, goby with spawn, whitebait, lobster
d. squid, mantis shrimp, plaice, sea bream, clam, bonito, mackerel, rainbow trout, conger eel
e. black rockfish, red sea bream, grunt, shrimp, bluefish, Spanish mackerel
f. freshwater shrimp, sharp-toothed eel, sea bass, bluefish, sweetfish
g. abalone, freshwater shrimp, crab, octopus, ray, grunt, eel, conger eel, sharp-toothed eel, flounder, sea bass
h. turbo, abalone, sweetfish, trout, loach, flounder, sea bass, sea bream
i. sea bream, sweetfish, jellyfish, conger eel, sea bass, sharp-toothed eel, sardine
j. mantis shrimp, mackerel, trout, grunt, hairtail
k. crab, squid, tiger shrimp, mackerel pike, tuna, yellowfish
l. fresh water clam, pond snail, sea urchin, sea cucumber, squid, puffer, yellowtail, tuna, salmon, gray mullet, wild boar, beef

Autumn is the season for cooking mackerel pike at home. With the first frost, one wants to check out the local grilled chicken stalls. This is when heavy catches of very yang fish such as yellowtail and tuna are made, and at no time are they more delicious. The exquisite flavor of yang fish during a yin season is certainly a part of nature's grand design. *Daikon* and leafy vegetables ready to gather from the garden go very well with these fish. People also know how to turn yin fish into yang food by salting or grilling, so meals are enjoyable and can be elevated to works of art.

Nothing surpasses the culinary artistry of home-made *miso* and *tofu* cooking and of fish cooked on the rocks by the river or at the fireside after flavoring with crude, natural salt prepared by burning sea salt with brush and seaweed.

And so with *osechi-ryori* dishes prepared for the New Year's. As cooking that celebrates the joy of the new year, the wisdom of pairing salted salmon, herring roe, kelp, and black soybeans, and adding sea bream and lobster goes beyond tradition to a perfect pairing of man and nature.

During the harsh cold of the winter months, mallard, jackrabbit, and other wild game served with welsh onions, leek, and wild rocambole warms the body. Even though food is scarce, the flavor of pickled vegetables gathered in the fall puts a fragrant finishing touch to a winter meal. And how can one describe the delightfully exotic taste of oysters, sea urchins, and sea cucumbers?

In late winter, on the verge of spring's arrival, the edible butterbur flower peeks through the cover of snow and the leaves of the creeping saxifrage beneath the snow are ready to be eaten. Hardy green herbs such as Japanese parsley, shepherd's purse, and chickweed can be found beneath the spring frost, and as one is appreciating the buds of the Japanese angelica tree, spring returns beneath one's window.

Spring comes quickly to Shikoku and by about the vernal equinox, field horsetails are emerging. This is a time for taking walks through fields of clover and picking the flowers. Some drink hot *sake* with their *sukiyaki* while others prefer sipping tea flavored with the petals of floating cherry blossoms.

In this way, the Japanese take foods of the seasons available near at hand, and while savoring well their excellent and distinctive flavors, are able to see the providence of the heavens in the modest fare on which they live. Within a quiet life passed leisurely and tranquilly according to the cycles of nature lies hidden all the grandeur of the human drama.

This farmer's diet, this diet of the fisherman on the coast who eats sardines with his potatoes and barley, these are also the common diets of the village people. Yes, they know what is delicious, but they have not neglected the subtle and curious flavors of nature.

A natural diet lies at our feet—a diet that obeys the laws of heaven and has been followed naturally and without want by the people of farming and fishing villages.

The Nature of Food

We normally think of food only as something needed by the body to live and grow, but what connection does food have with the human soul?

For animals, it is enough to eat, play, and sleep. Nothing could be better than if man too were able to live a life of contentment enjoying nourishing food, health, and tranquil sleep. What does it mean to enjoy and take pleasure in food? This, along with nourishment and nutrition, is a question of both matter and spirit.

Buddha said, "Form is emptiness and emptiness is form." Since "form" in Buddhist terminology refers to matter and "emptiness" to spirit, matter and spirit are one. Matter has many aspects, such as color, shape, and quality, each of which affect the spirit in many ways. This is what is meant by the unity of matter and spirit.

Chief among the aspects of matter serving as food are color and flavor.

Color*:   The world appears to be filled with seven basic colors, but when combined these seven colors become white. In a sense, one could say that what was originally white light was divided into seven colors with a prism. Viewed with detachment, all things are colorless and white. But to one distracted, seven moods (spirit) engender seven colors (matter). Matter is spirit and spirit is matter. Both are one.

Water undergoes countless transformations but remains always water. In the same way, beneath the infinite variety of creation, all things are essentially one; all things have basically one form. There was never any need for man to categorize everything. Although differences may exist between the seven colors, they are all of equal value. To be distracted by these seven colors is to fail to note the matter and spirit underlying them, to be sidetracked by the inconsequential.

The same is true of food. Nature provided man with a vast array of foods. Discerning what he thought were good and bad qualities, he picked and chose, thinking that he had to create harmonious combinations and blends of color, that he should always partake of a rich variety. This has been the root of his mistake. Human knowledge can never compare with the greatness of the natural order.

We have seen that there never was an east or west in nature; that left and right, yin and yang did not exist; that the Right Path, the path of moderation as seen by man, is not that at all. People may say that there is yin and yang, that seven colors exist in nature, but these are only products of the entanglement of the labile human spirit and matter; they change constantly with time and circumstance.

The colors of nature remain constant and immutable, but to man they appear to change as readily as hydrangea blossoms. Nature may seem ever-changing but because this motion is cyclical and eternal, nature is in a sense fixed and immobile. The moment that man halts the seasonal cycle of foods on whatever pretext, nature will be ruined.

---

* The Chinese character for color (色) is used in Buddhist texts to represent form or matter.

The purpose of a natural diet is not to create learned individuals who support their selection of foods with articulate explanations, but to create unlearned people who gather food without deliberate rationale from nature's garden, people who do not turn their backs on Heaven but accept its ways as their own.

A true diet begins through detachment from shades of color, by delighting in colors without hue as true color.

Flavor: People will say, "You can't know what something tastes like unless you try it." Yet a food may taste good or bad depending on when and where it is eaten. Ask the scientist what flavor is and how one comes to know a flavor, and he will immediately begin analyzing the ingredients of the food and investigating correlations between the minerals extracted and the five tastes—sweet, sour, bitter, salty, and hot. But flavor cannot be understood by relying on the results of a chemical analysis or the sensations at the tip of the tongue.

Even were the five tastes perceived by five different organs, a person would be unable to sense the true flavor if his instincts themselves were confused. Scientists may extract minerals and study the movement of the heart and the physical response following sensations of deliciousness and pleasure, but they do not know what makes up the emotions of joy and sorrow. This is not a problem that can be solved with a computer. The physician thinks that an investigation of the brain cells will give the answer, but a computer programmed to think that sweet is delicious is not likely to feed out the result that sour is delicious.

Instinct does not investigate instinct, wisdom does not turn back and scrutinize itself. Studying how the seven flavors of the seven herbs of spring act upon the human sense of taste is not what is important. What we must consider is why man today has parted with his instincts and no longer seeks to gather and eat the seven herbs of spring, why his eyes, ears, and mouth no longer function as they should. Our primary concern should be whether our eyes have lost the ability to apprehend real beauty, our ears to capture rare tones, our nose to sense exalted fragrances, our tongue to distinguish exquisite tastes, and our heart to discern and speak the truth. Flavors caught with a confused heart and numbed instincts are a far cry from their true selves.

Evidence that the human sense of flavor has gone haywire is difficult to find, but one thing is certain: people today chase after flavor because they have lost it. If this sense were intact, they would be able to judge accurately for themselves. Even though natural man gathers his food without discrimination, his instincts are intact, so he eats properly in accordance with natural laws; everything is delicious, nourishing, and therapeutic. Modern man, on the other hand, bases his judgments on mistaken knowledge and searches about for many things with his five deranged senses. His diet is chaotic, the gap between his likes and dislikes deepens, and he hurdles on toward an even more unbalanced diet, drawing his natural instincts further away from true flavor. Delicious food becomes increasingly rare. Fancy cooking and flavoring just compound the confusion.

The problem then, as I can see, is that man has become spiritually alienated from food. True flavor can be perceived only with the five senses, the mind, and the spirit. Flavor must be in consonance with the spirit. People who think that flavor originates in the food itself eat only with the tip of the tongue and so are easily deceived by the flavor of instant cooking.

An adult who has lost his instinctive sense of taste no longer appreciates the taste of rice. He normally eats white rice prepared by polishing brown rice to remove the bran. To make up for the loss in flavor, he adds meat sauce to the white rice or eats it together with *sashimi*. Tasty rice thus becomes rice that is easy to flavor and season, and people delude themselves into thinking of white rice, which has been stripped of the aroma and taste peculiar to rice, as high-grade rice. I imagine that some people think it better to eat enriched rice than to try and squeeze any nutrition out of highly polished rice, or they rely on side dishes of meat or fish for the necessary nutrients. Nowadays it is all too easy to believe that protein is protein and vitamin B is vitamin B regardless of where these come from.

But through a major lapse in thinking and responsibility, meat and fish have gone the same route as rice. Meat is no longer meat and fish no longer fish. Refinements in flavoring with petroleum-derived protein have created people unaware and unconcerned that their entire diet has been converted into an artificial diet.

Today, the locus of flavor is the food product. Thus beef and chicken are "delicious." But it is not eating something "delicious" that satisfies the palate. All the conditions must be right for something to be sensed as delicious. Even beef and chicken are not delicious per se. The proof is that to people who have a physical or mental aversion to meat, these are unpalatable.

Children are happy because they are happy; they can be happy playing or doing nothing. Even when adults are not especially happy but they believe they are enjoying themselves, as when they watch television or go to see a baseball game, they may gradually become happy and even break out laughing. Similarly, by removing the original conditions that planted the idea in someone's head that something is unappetizing, this can become delicious.

There is a folk tale in Japan of how, deceived by a fox, someone was made to eat horse manure. But it is not for us to laugh, for people today eat with their minds and not with their body. When they eat bread, it is not the flavor of the bread they enjoy, but the flavor of the seasonings added to the bread.

People nowadays seem to live by feeding on a mist of notions. Man originally ate because he was alive, because something was delicious, but modern man eats to live and thinks that if he does not prepare and dine on choice cuisine, he will not be able to eat delicious food. Although we should pay more attention to creating individuals who can enjoy eating anything, we have put aside the person and spent all our efforts on preparing delicious food. This has had the opposite effect of reducing the amount of delicious food we eat.

In our efforts to make bread tastier, bread has ceased to taste good. We have grown energy-extravagant crops, livestock, and fowl to create a world of plenty, and instead triggered famine and starvation. What foolishness, all of this. But man's inability to recognize the folly inherent in his efforts has thrown him into greater confusion. Why is it that the more he strives to produce delicious rice, fruit, and vegetables, the more inaccessible these become? I often run into people who are perplexed as to why delicious food can no longer be found in Tokyo.

They fail to notice that man's efforts to set up all the conditions for producing delicious rice or apples have distanced him from true flavor. Unfortunate as it may be, city dwellers have lost a true sense of taste. Everyone works so hard to make something delicious that they end up by deceiving themselves into thinking it so. No one attempts to look directly at the truth of flavor. The only ones that win out are the manufacturers that exploit these deceptions and the merchants who hop a ride to make a buck.

What does it take to come by truly delicious food? All we have to do is stop trying to create delicious food and we will be surrounded by it. However this will not be easy as cooking and cuisine are regarded as worthwhile and essential activities—part of the culture of food. Ultimately, true cooking and the pursuit of true flavor are to be found in a comprehension of the subtle and exquisite flavors of nature.

People today who cannot eat wild herbs without removing their natural astringency are unable to enjoy the flavors of nature. The practical wisdom of early man who sun-dried root vegetables and pickled them in salt, rice bran, or *miso*, enjoying their special taste and aroma at the end of his meals; the delicious flavor and nourishment of food cooked with salt; the subtle and singular flavors created from an existence that relied on a single kitchen knife; ... these are understood by everyone everywhere because they touch the essence of the flavors of nature.

Long ago, people of the aristocratic classes in Japan used to play a game called *bunko* (聞香) in which players had to guess the fragrances of various types of burned incense. It is said that when the nose was no longer able to distinguish the aromas, the player bit into a *daikon* root to restore the sense of smell. I can just imagine the expression on the face of an aristocrat chomping into a length of pungent *daikon*. This shows plainly that taste and aroma are exuded by nature.

If the purpose of cooking is to delight people by modifying nature in order to bring out an exotic flavor that resembles nature but is unlike anything in nature, then we are dealing with deceit.

Like a sword, the kitchen knife may do good or evil, depending on the circumstances and who wields it. Zen and food are one. For those who would sample the delights of a natural diet, there is Buddhist vegetarian cooking and Japanese high tea. An unnatural afternoon tea may be served in high-class restaurants to which farmers shod in workboots are not welcome, but modest, natural teas have disappeared. When coarse green tea sipped by the open hearth is more delicious than the refined green tea of the tea ceremony, this spells an end to the tea culture.

Culture is seen as a human product created, maintained, and refined through human invention by the separation of man from nature. However, the culture actually associated intimately with daily existence and handed down and preserved to later generations always originates in a return to the source of nature (God), forming of itself when nature and man fuse into a single whole. A culture born of human recreation and vanity that is divorced from nature cannot become a true culture. True culture arises from within nature and is pure, modest, and simple. Were this not so, then man would surely be destroyed by that culture. When mankind forsakes a natural diet for a civilized diet, he turns away from a true culture and sets off on the road to decline.

I noted above that the knife which the cook wields is a two-edged sword. It can lead to the way of Zen. But because diet is life, a diet that strays from the true principles of nature robs man of his life and sends him down the wrong path.

The Staff of Life: Nothing is better than eating delicious food, but how often do we hear that food is eaten to support the body and draw nourishment? Mothers are always telling their children to eat their food, even if they do not like it, because it is "good" for them. Here we have another example of a reversal in human thinking. This is the same as saying that we nourish ourselves so that we can work harder and live longer.

Taste and nutrition should not be separated. What is nourishing and good for the human body should stimulate the human appetite of its own accord and serve as delicious food. Flavor and nutrition must be one.

Not so long ago, farmers in this area enjoyed simple meals of barley and rice with unrefined soy sauce and pickled vegetables. This gave them strength and long life. Stewed vegetables and rice cooked with adzuki beans was a once-a-month treat. How was this enough to supply their nutrient needs? Rather than thinking in terms of "drawing nutrition," it makes more sense to say that working hard in the fields made one hungry, which is why coarse fare tasted delicious. And, of course, a strong body can draw sustenance from a simple diet.

In contrast with the simple Eastern diet of brown rice and vegetable that provides everything the body needs, Western dietetics teaches that health cannot be maintained unless one has a balanced diet with a full complement of nutrients: starch, fat, protein, vitamins, minerals, and so on. It is no surprise then that some mothers stuff "nutritious food" into their children's mouths, regardless of whether it tastes good or not.

Because dietetics is built upon careful scientific reasoning and calculation, the general tendency is to accept its pronouncements at face value. But this carries with it the potential for disaster.

First of all, dietetics lacks any awareness of man as a living, breathing creature. Menus leave the impression that one is merely supplying energy to mechanical humans cut off from the source of life. There is no evidence of any attempt to approach closer to a natural existence, to conform to natural cycles. In fact, because it relies so heavily on the human intellect, dietetics appears useful rather in the development of anti-natural man isolated from nature.

Secondly, it seems almost as if we had forgotten that man is a spiritual animal that cannot be fully explained in organic, mechanical, or physiological terms. He is an animal whose body and life are extremely fluid and which undergoes great physical and mental vicissitudes. Things might be different if there were guinea pigs that could speak, but there are limits to how far scientists can go in extrapolating the results of dietetic experiments on monkeys and mice to man. The food that man eats is linked directly and indirectly with human emotion, so a diet devoid of feelings is meaningless.

Third, Western dietetics understands things only within a narrow temporal and spatial framework; it cannot grasp things in their entirety. No matter how the scientist may attempt to assemble a full array of ingredients, this will never approach a complete diet. The powers of the intellect will succeed only in the creation of an incomplete diet far removed from nature. Unmindful of the simple truth that "the whole is greater than the parts," modern science commits blunder after blunder. Man can dissect a butterfly and examine it in the greatest detail, but he cannot make it fly. And even were this possible, he could not know the heart of the butterfly.

Let us look at what goes into the preparation of a daily menu in Western fashion. Naturally, it will not do to eat randomly anything that comes one's way. A daily menu is normally drawn up by thinking of what and how much one should eat each day to achieve a balanced diet. I would like to take as my example the four-group scoring method used at the Kagawa Nutrition College in Japan. Here are the four groups with the type of food they represent and the number of points allotted daily to each.

Group 1: Good protein, fat, calcium, and vitamin foods such as milk and eggs for complete nutrition—3 points.

Group 2: Bluefish, chicken, and *tofu* as nutrients for building muscle and blood—3 points.

Group 3: Light-colored vegetables, green and yellow vegetables, potato and mandarin oranges to provide vitamins, minerals, and fiber for a healthy body—3 points.

Group 4: White rice, bread, sugar, and oils as sources of sugar, protein, and fat for energy and body temperature—11 points.

Since each point represents 80 calories, a day of balanced meals gives 1,600 calories. Because it provides 80 calories, 80 grams of beef is worth one point, as is 500 grams of bean sprouts, 200 grams of mandarin orange, and 120 grams of grapes. Eating 40 oranges or 20 bunches of grapes each day would give the necessary calories but would not make for a balanced diet, so the idea here is to eat a mixture of foods from all four groups.

This appears to be eminently sensible and safe, but what happens when such a system is employed uniformly on a large scale? A year-round supply of high-grade meat, eggs, milk, bread, vegetables and other foods has to be kept ready, which necessitates mass production and long-term storage. This just might be the reason why farmers have to grow lettuce, cucumbers, eggplant, and tomatoes in the winter.

No doubt, the day is not far off when farmers will be told to ship out milk in the winter, mandarin oranges in early summer, persimmons in the spring, and peaches in the fall. Can we really have a balanced diet by gathering together many different foods at all times of the year, as if there were no seasons? The plants of the mountains and streams always grow and mature while maintaining the best possible nutrient balance. Out-of-season vegetables and fruit are unnatural and incomplete. The eggplants, tomatoes, and cucumbers grown by natural farming methods under the open sun ten or twenty years ago are no longer to be found. Without a distinct fall or winter it is hardly surprising that the eggplants and tomatoes produced in greenhouses no longer have the flavor or fragrance they used to. One should not expect these to be packed with vitamins and minerals.

Scientists see themselves as working to ensure that people get all the nourishment they need anywhere and anytime, but this is having the opposite effect of making it increasingly difficult to obtain anything but incomplete nourishment. Nutritionists are unable to grasp the root cause of this contradiction for they do not suspect that the first cause for error lies in the analysis of nutrition and the combination of different nutrients.

According to the principle of yin and yang, the basic foods listed above such as meat, milk, chicken, and bluefish are highly yang and acidic, while potato is a very yin vegetable. None of these agree with the Japanese people. This then is the worst possible list of foods.

Today in Japan we have more rice than we know what to do with and barley is being phased out, but if we grew rice suited to the climate of this "Land of Ripening Grain," stopped importing wheat, grew early-maturing naked barley that can be harvested during May before the early summer rains, and revived the practices of eating brown rice and rice-and-barley like the farmers and samurai of old—if we did all these things, then we would see an immediate improvement in Japan's food situation and the health of her people. If all this is asking too much of modern man, with his weakened heart and stomach, then I would recommend at least that he make brown rice bread or delicious bread from naked barley.

Farmers too give no thought to the meaning of a natural diet or natural farming, and without a trace of skepticism see the production of food out of season as a method for increasing the food supply. Scientists and engineers follow suit, working on the development of new food products and research on new methods of food production. Politicians and those in the distribution industry believe that markets well stocked with a full range of goods means food is abundant and people can live in peace and security, but such thinking and the follies of people are dragging mankind to the abyss of destruction.

Summing Up Natural Diet

There are four major types of diet in this world:

1. A lax, self-indulgent diet, influenced by the external world, that submits to cravings and fancies. This diet, directed by the mind, might be called an empty diet.
2. The physically centered diet of most people, where nutritional food is consumed to sustain the body. This is a scientific diet that spins centrifugally outward with increasing desires.
3. The diet of natural man based on spiritual laws. Extending beyond Western science and centered on Eastern philosophy, it places restrictions on foods, aiming for centripetal convergence. This could be called a diet of principle and includes what is normally referred to as "natural diet."
4. A diet that lays aside all human knowledge and by which one eats without discrimination in accordance with divine will. This is the ideal natural diet and constitutes what I call a "non-discriminating diet."

258

People should begin by discarding empty, self-indulgent diets that are the root of a thousand diseases, and, failing to find satisfaction in a scientific diet that does no more than sustain the life of the organism, move on to a diet of principle. But they must then go beyond theory and strive towards the ultimate goal of becoming true people that partake of an ideal natural diet.

The Diet of Non-Discrimination: This is founded on the view that man does not live through his own efforts but was created and is supported by nature.

The diet of true man is life and sustenance provided by the heavens. Food is not something that man selects from within nature, it is a gift bestowed upon him from above. Its character as food lies neither exclusively in itself nor in man. A true natural diet becomes possible only when food, the body, and the soul fuse together completely within nature. What could be called a diet of non-discrimination achieved by the union of nature and man is a diet that the self, which is infused with and embodies the will of heaven, takes subconsciously.

True man with a truly healthy body and mind should be naturally equipped with the ability to take the right food from nature, without discrimination or error.

To follow the will of the body and desire freely, to eat when a food is delicious and forbear when it is not, to partake without restraint, without plan or intention, is to enjoy the most subtle and exquisite fare—an ideal diet.

Ordinary man must work toward the ultimate goal of an ideal natural diet by first practicing a natural diet that falls one step short of this ideal and striving earnestly to become natural man.

The Diet of Principle: All things exist in nature. Nothing is lacking; nothing is present in excess. The foods of nature are complete and whole in and of themselves. It should always be remembered that nature too is a single, harmonious whole, ever complete and perfect.

It is only fitting that nature is not subject to man's criteria, to his choosing and rejecting, his cooking and combining. Man thinks that he can explain and expound on the origin and order of the universe, on the cycles of nature. It appears as if, by applying the principle of yin and yang, he can achieve harmony of the human body. But if, ignorant of their limits, he becomes caught up in these laws and tenets and uses human knowledge indiscriminately, he commits the absurdity of looking closely at the small and insignificant without catching sight of the larger picture, and of taking a broad view of nature while failing to notice the details at his feet.

Man can never understand one part of nature, much less the whole. Mankind may think itself the orphan of the natural world, but the position taken by those who long earnestly for a natural diet is to renounce human knowledge and submit to the will of nature by reaffirming one's obeisance to divine providence. It is already enough to eat cooked and salted food, to consume all things in moderation, to gather foods of the seasons that grow close at hand. What one must then do is to devote oneself fully to the principles of wholism, the inseparability of the body from the land, and a simple local diet. People must realize that a diet of surfeit which relies on foods from far-off lands leads the world astray and invites human ills.

The Diet of the Sick: A natural diet appears irrelevant, primitive, and crude to people who practice an empty diet of self-indulgence in pursuit of flavor and to those who think of food only as matter needed for sustaining biological life. But once they realize that they are in poor health, even they will begin to show an interest in natural diet.

Illness begins when man moves away from nature, and the severity with which he is afflicted is proportional to his estrangement. This is why if a sick person returns to nature he is cured. As mankind distances itself from nature, the number of sick people rises rapidly and desires for a return to nature intensify. But attempts to return to nature are thwarted because people do not know what nature is, nor do they know what a natural body is.

Living a primitive life deep in the mountains, one may learn what non-intervention is but will not know nature. Yet taking some action is also unnatural.

Lately, many people living in the cities have been trying to obtain natural food, but even if they succeed, without a natural body and spirit prepared to receive such food, merely consuming it does not constitute a natural diet.

Farmers today are simply not producing natural foods. Even if urban people wished to establish a natural diet, no materials are available. Moreover, it would probably take almost superhuman skills and judgment to live on a complete natural diet in a city under such conditions and eat meals with a yin-yang balance. Far from returning to nature, the very complexity of eating a natural diet in this way would just drive people further away from nature.

To push upon people living in different environments and of different types and temperaments a rigid, standardized natural diet is an impossibility. This does not mean, however, that various types of natural diet exist. Yet just look at the different natural diet movements being espoused around the world.

One such movement claims that because man is basically an animal he should eat only uncooked food. A few say that man should drink broth prepared from raw leaves, while some physicians warn that following a raw diet without knowing fully what one is doing is dangerous. There are natural diets based on brown rice and scientists who proclaim the merits of white rice. Some claim that cooking food enriches the human diet and is good for the health, while others argue that this only helps to create sick people. To some, fresh water is good; to others, bad. Some acclaim salt to be invaluable while others attribute a whole range of diseases to excessive salt intake. One camp sees fruits as yin and food fitting for monkeys perhaps but not for man while another contends that fruit and vegetables are the best possible foods for health and longevity.

Given the right circumstances, any one of these views is correct, so people end up thoroughly confused by what appear to be so many contradictory claims. Nature is a fluid entity that changes from moment to moment. Man is unable to grasp the essence of something because the true form of nature leaves nowhere to be grasped. People become perplexed when bound by theories that try to freeze a fluid nature. One misses the mark if he relies on something that is unreliable.

Right and left do not exist in nature, so there is no happy medium, no good and evil, no yin and yang. Nature has given humanity no standards to rely upon.

It is senseless to arbitrarily decide, independent of the land and the people, what the food staple and the minor foods should be. This just moves one further away from true nature.

Man does not know nature; he is like a blind man without any idea of where he is headed. He has had no choice than to take science's cane of knowledge and tap out the road at his feet, relying on the principle of yin and yang to set the direction of his travels, like the stars in the night sky. Whatever direction he has taken, he has thought with his head and eaten with his mouth. What I wish to say is that he must stop eating with his head and clear his mind and heart.

The food mandalas I have drawn (Figs. 5.4 and 5.5) are more valuable than the most lengthy discourse. I meant these to be used as a compass by which to set one's course, according to the circumstances and the degree of sickness or health, either for a centrifugal diet or a centripetal diet. But once these mandalas have been examined, they may be discarded. By this I mean that people should not eat on the basis of human intellect and action, but should merely receive with gratitude the food that grows in nature.

Before this can be done, however, people must first become natural people and the ability of the body to select foods and properly digest them must be restored. If natural people arise who, instead of following a natural diet that prescribes this and proscribes that, are satisfied without anything, then everything will be resolved. Rather than pursuing a natural diet that cures the sick, the first priority should be to return to nature and to a healthy natural man. Those very people normally thought of as healthy I would call the seriously ill; saving them is of the greatest importance. Doctors are busy saving sick people, but no one is reaching out to save the healthy. Only nature itself can do so. The greatest role of a natural diet is to return people to the bosom of nature. The young people living primitively in the orchard huts on the mountain, eating a natural diet and practicing natural farming, stand closest to the ultimate goal of mankind.

Conclusion: Natural farming, natural diet, and natural healing are all part of one whole. Without an established natural diet, farmers have no idea what it is they should produce. Yet nothing is clearer than that, in the absence of an established method of natural farming, a true natural diet will never take hold and spread. Both natural diet and natural farming can be achieved only by natural people. This trinity begins and is realized at once. The goal of all three elements is the creation of ideal man.

However, man's ideals today are in a state of confusion; a hundred schools of thought on natural diet and natural farming compete for our attention. The bookstores are flooded with books on natural diet, and magazines and journals are full of articles on organic farming, microbial farming, enzyme farming, and other methods that depart from scientific farming. But to me, these all look pretty much the same. They are all on the same level and amount to no more than just one field of scientific agriculture.

People look on complacently, thinking that the world goes on developing in the

midst of repeated chaos and confusion, but fragmented development without
a goal can lead only to chaotic thinking and, ultimately, destruction of the human
race. Unless we succeed very soon in clarifying what nature is and what man
should and should not do, there will be no turning back.

3. Farming for All

Advances in modern civilization appear to have made our lives easier and more
convenient. Life in Japan's large cities has reached about the same level of affluence
as in advanced Western countries, and the youth who glorify freedom seem to be
easy at heart. But all that has really grown is the economy. The inner life of people has become stunted, natural joy has been lost. More and more people have
turned to standardized forms of recreation such as television, the *pachinko* parlor,
and mah-jongg, or seek temporary solace through drinking and sex.

People no longer tread over the bare earth. Their hands have drawn away from
the grasses and flowers, they do not gaze up into the heavens, their ears are deaf
to the songs of the birds, their noses are rendered insensitive by exhaust fumes,
and their tongues have forgotten the simple tastes of nature. All five senses have
grown isolated from nature. People have become two or three stages removed
from true man in the same way that someone riding in his car over asphalt-paved
roads is two or three stages removed from the bare earth.

Progress in Japan since the Meiji Reformation has brought material confusion
and spiritual devastation. Japan can be likened to a patient dying of cultural
disease who is submitted to a medical experiment. This condition is the fruit of
the "cultural flowering" to which all of Japan applied itself throughout the Meiji,
Taisho, and Showa eras. We must call a halt to this flowering of destruction now.
The objective of my "do-nothing" philosophy is the revival of villages of true
man where people can return to the original form of nature and enjoy genuine
happiness. The program to achieve this I shall refer to simply as "Farming for
All."

Creating True People

False materialistic culture and agriculture begin and end by "doing," but the way
of true man begins and ends by "doing nothing."

The road of true man is an inner road. It cannot be followed by advancing
outwards. We can unearth the precious kernel of truth that lies buried within
each of us by first throwing off the delusions in which we are attired.

The path of a "do-nothing" nature where all one does is to plunge into the
bosom of nature, shedding body and mind, this is the road that true man must
walk. The shortest path to attaining the state of true man is an open existence

with simple garments and a simple diet, praying down to the earth and up to the heavens.

True and free happiness comes by being ordinary, it is to be found only by following the extraordinary, methodless road of the farmer, irrespective of the age or direction. Spiritual development and resurrection are not possible if one strays from this road of humanity.

In a sense, farming was the simplest and also the grandest work allowed of man. There was nothing else for him to do and nothing else that he should have done.

Man's true joy and delight was natural ecstasy. This exists only in nature and vanishes away from the earth. A human environment cannot exist apart from nature, and so agriculture must be made the foundation for living. The return of all people to the country to farm and create villages of true men is the road to the creation of ideal towns, ideal societies, and ideal states.

The earth is not just soil, and the blue sky is more than just empty space. The earth is the garden of God, and the sky is where He sits. The farmer who, chewing well the grain harvested from the Lord's garden, raises his face to the heavens in gratitude, lives the best and most perfect life possible.

My vision of a world of farmers is founded on the responsibility of all people to return to the garden of God to farm and their right to look up at the blue skies and be blessed with joy. This would be more than just a return to primitive society. It would be a way of life in which one constantly reaffirms the source of life ('life' being another name for God). Man must also turn away from a world of expansion and extinction, and place his faith instead in contraction and revival.

This society of farmers may of course take the form of peasant farming, but it must comprise natural farming that transcends the age and searches earnestly for the wellsprings of agriculture.

The Road Back to Farming

Recently, led by individuals aware of the danger of being swallowed up by urban civilization, people in the great metropolises, cut off from the natural world, have sensed within themselves a heightened need for nature and have even begun seeking a road back to farming. What is it that keeps them from realizing their dreams but themselves, the land, and the law? Do people really love nature? Do they really intend to return to the land and here build up a society where they may live in peace and comfort? Somehow, it does not look that way to me.

Even when I think that the hopes and views of these people are absolutely correct, I cannot help feeling a sense of futility and distance in the end. It is something like scooping up duckweed floating on the surface of a pond and watching it slip through one's fingers. There do not seem to be any links between people, between man and nature, between up and down, right and left.

Although both encounter the same nature, the city youth sees a natural world that is nothing more than a vision or dream, while what the rural youth works is not earth, but merely soil. Between the producer and consumer, both of whom are concerned with the same problems and should handle these jointly, lies an endless parade of organizations, merchants, and politicians. Superficial connections exist between these, but one can sense the internal gaps, the misery of those who share a common task but different dreams, the impatience of those who float on the same waves but do not notice that they drink the same water.

The consumer, who denounces food contamination, has himself sown the seeds of pollution. He does not think it strange that agricultural science has fluorished and the farmer declined. The politician who laments over the course taken by modern agriculture rejoices at the decrease in the number of farmers; corporations that have prospered from an agricultural base have brought farmers to ruin.

Farmers themselves have destroyed the earth while praying for its protection. People attack the destruction of nature yet condone destruction in the name of development. They make compromises in the name of harmony and prepare for the next wild rampage.

The foremost cause for the discord and contradictions of human society is that everyone in the towns and cities act independently and in their own interest without seeing things clearly. People all claim to love nature, but each pushes his interests without feeling the least contradiction or concern.

The lack of coherence in this world and the flood of disjointed campaigns attest to one thing: what everyone really loves is not nature but himself. The painter who sketches the mountains and rivers appears to love nature, but his real love is sketching nature. The farmer who works the earth merely loves the image of himself laboring in the fields. The agricultural scientist and administrator believe they love nature, but the one only really loves the study of nature and the other enjoys studying and passing judgment on the farmers at work. Man has glimpsed but one tiny portion of nature. People only think they understand its true essence; they only think they love nature.

Some people transplant trees from the mountain into their garden as a token of their love for nature while others plant trees in the mountains. Some say that going to the mountains is faster than planting trees or demand that roads be laid down to make the mountains more accessible while others insist on walking to the mountains rather than going by car. All wish to adore nature but by different means so they believe that the only solution is to advance while maintaining harmony in some way. However, because their perception and understanding of nature is superficial, these methods of appreciating nature are all at odds with each other. If each and every individual penetrated to the very core of nature and truly understood its essence, then no difference of opinion could arise.

No "method" is needed for loving nature. The only road to nature is non-action, the only method is no method at all. All one must do is to do nothing at all. The means will become clear of itself and the goal absurdly easy to attain.

This is what I mean by doubting the degree of resolution in those who profess a wish to return to nature. Are they really drawn to farming? Do they really love nature? If you have a genuine love for nature and wish to return to farming, the way will open with great ease before you. But if your love for nature is superficial and what you do amounts simply to making use of farming for your own purposes, the road will be closed off to you; returning to nature will be impossibly difficult.

The obstacle that blocks the first step back to the land is people; it lies within yourself.

Enough Land for All

The second obstacle blocking the return of people to the land is the availability of farmland. With 120 million people squeezed together in a small island nation and land prices soaring out of sight, purchasing farmland would appear to be next to impossible. I have chosen nevertheless to call my program "Farming for All."

Japan has about 15 million acres of farming land, which works out to about a quarter-acre per adult. If Japan's land were divided into 20 million households, this would give each household three quarters of an acre of farmland plus two and a half acres of mountain and meadow land. With total reliance on natural farming, all it takes to support a household of several people is a quarter-acre. On this amount of land, one could build a small house, grow grains and vegetables, raise a goat, and even keep several chickens and a hive of bees.

If everyone were capable of being content with the life of a quarter-acre farmer, then this would not be impossible to achieve. More to the point, everyone has a right and a duty to live their lives within narrow bounds. This is the basic condition for achieving an ideal life.

People, feeling shackled by laws and stratospheric land prices, view the possibility of land ownership as hopeless, but there is plenty of land to be had. Laws exist basically to protect an ideal society. Why then have land prices risen to such dizzying heights out of the reach of the nation's people?

Rises in land prices over the past several years were triggered first by massive purchases of land for housing and public use. This arose both from a general perception, aided by publicity to this effect, that land in Japan is in short supply—a limited resource that cannot be increased, and the convergence of people, drawn by false rumors of economic growth, on the cities. But the truth is that, no matter how much the population grows in Japan, there will always be more than enough land to build houses on. There is land aplenty, but that land classified as "housing land" has become a life-threatening cancer.

The law breaks the land up into different zones according to use: forests, farmland, housing land, and so forth. The City Planning Law was enacted, based upon which lines were drawn and farmland divided into areas lying within urban planning zones, areas within land adjustment zones, and areas outside of the lines. The conversion of farmland to housing land was prohibited. This forced a sharp reduction in housing land that pushed up prices. Enforcement of the National Land Use Planning Law may have made land easy to come by for those enforcing it, but it made land even more inaccessible to the ordinary person.

As laws proliferate, they appear to move toward perfection, but they only become more imperfect and preposterously complex, breaking man and the land apart. Only those who know the law well and can change the official category of a piece of land are able to buy land and later sell it. Every time housing land changes hands, the price goes up. If it were possible, just for the sake of argument, for anyone to build a simple hut or house anywhere he pleases without any legal formalities, then there would be an essentially unlimited supply of housing land. But for some reason, lawyers and legislators are under the impression that such a house would not be an ideal house.

So many legal constraints exist to building a house satisfying the legal definition of a house that the house cannot be built. A mountain hut or shed such as a woodcutter or farmer might use for his work is permitted, but if someone were to erect a small house in which he laid *tatami* mats, hung a lantern, and laid some water pipes, the land on which it was built would have to be housing land. But land classified as housing land must be serviced with a 13-foot-wide road and plumbing for tap water and sewage. Thus the prospective homeowner has no choice but to buy land developed for housing at a high price from a real estate firm and build a costly house that satisfies all the standards and codes. This system of legal stipulations has set into motion a negative cycle that has shot prices for housing land out of sight. Unscrupulous business practices taking advantage of the situation have further complicated the housing land problem, causing prices to jump even higher and driving people who want a house and land into a state of frenzy.

This also makes it difficult for people with aspirations of becoming quarter-acre farmers to purchase farmland. It is not that there is no farmland available, but that there is no category of land that anyone can freely work. There is no need to go into a sparsely populated mountainous area to find an example. Not one square yard of land that goes by the name of farmland is available for purchase by city people. Such land cannot be bought by anyone other than a farmer. Legally, a "farmer" is someone who owns at least 1-1/4 acre of farmland. The Agricultural Land Law has brought a halt to the transfer of farmland.

Unless someone from the city buys up at least 1-1/4 acre of land at once, he cannot become a farmer. In fact, non-farmers can neither buy land nor formally lease and work it as tenant farmers. But there are always loopholes in the law. For example, if earth is carried onto a piece of farmland or if the land is gradually turned into a lumberyard or flowers and trees planted, then with time it can be converted into a category called "miscellaneous land." Once this is done, the land can easily be sold or a house built on it. Even so, in sparsely populated areas, unused land is left abandoned because it cannot be transferred or leased for the simple reason that the use category cannot be changed.

The mountains, forests, and other wilderness land that accounts for about eighty percent of Japan's land area is tied up by titles and laws that prevent its practical use. If even a small portion of this area were freed for use as agricultural land, homesteading would begin immediately. These farmlands can be expanded and made fluid not by establishing new laws but by abolishing unnecessary ones. Laws that do not arise and are not consummated naturally do not remain in force very long.

The current price of farmland has been artificially inflated over the natural price. Until recently, the price of farmland has always been stable, remaining more or less fixed at a given level. For prime farmland, the best price was 110 bushels of rice per quarter-acre. Assuming a bushel of rice to cost $20, this comes to $2,200 per quarter-acre. Figuring that whoever bought the land could not make ends meet if prices went any higher, farmers used this rate as a yardstick whenever they bought and sold land to each other. This standard should continue to be maintained.

Prices and taxes on farmland became unjustly high when it began to be assessed on the same scale as housing land by local government. This was clearly designed to drive farmers off the land by burdening them with taxes too high for them to afford with the meager earnings generated on their land. Support was easily drummed up among city dwellers with the argument that once farmland was freed for use as housing land, the increase in housing land would probably drive down prices. But this turned out to be just wishful thinking; land released in this way never came within the reach of the common man. The oases of green left in the towns and cities are no longer farmlands and are vanishing from the reach of the farmer. This tragedy will surely come to trouble all the farmers of this land. Someday too, these hardships of the farmer will return, in the form of calamity, to threaten the well-being of those living in the cities.

The problem boils down to this: only scoundrels, the clever, and those in power stand to gain from the issuance and abuse of a barrage of capricious laws. The net result is that the land is taken out of the hands of farmers. The Agricultural Land Law, established to protect tenant farmers, today serves no other purpose than to thwart the hopes of those wishing to become farmers.

No one knows more about farmland than farmers. If things had been left in their hands, there would have been no need for any laws. The farmer would have passed the land into the hands of his children or grandchildren when the time came to do so. If for some reason it became necessary to transfer ownership of the land, the farmer would have resigned himself to the inevitable and passed his fields into the hands of his neighbor smoothly and without the least trouble.

When people can do without a law, it is best not to have that law. Only the barest minimum of laws is needed—to create a world that can get along without laws. Were it necessary to have a single law, then it should be: "One shall build one's house at least sixty feet from one's neighbor." If people were to scatter out and build a small house on a quarter-acre wherever they pleased, then the food problem would take care of itself, water and sewage lines would not be needed, and the problem of pollution would be licked. That is not all; this would also be the quickest path to making our lands a paradise on earth.

It is not that land is not available for housing and farming. For individuals burning with a desire to farm the open land and willing to learn some basic skills, farmland exists everywhere. There is no limit to the places one can live.

Running a Farm

Even if aspirants to farming are able to purchase land, what are the chances that they will be able to support themselves? Up until a few decades ago, seventy to eighty percent of the Japanese people were small farmers. Poor landed peasants were called "one-acre farmers." If peasants were barely able to eek out a living on one acre, then what hope is there for someone planning to live off a quarter-acre?

But the reason the farmers of the past were poor and hungry was not that their land was too small to support them. Their poverty was not of their own making. They were the victims of outside forces: an oppressive social system and political and economic mechanisms beyond their control.

A quarter-acre of land is enough to supply the food needed for supporting a family. If anything, a full acre is too large. Had the peasants been exuberant at heart and ruled over by a benevolent government, instead of living in mean poverty, they could have lived like princes on their acre of land.

Farmers at the time were said to grow a hundred crops. In the paddy field and vegetable gardens, they raised rice, barley, and other grains, as well as sweet potato and many different vegetables. Fruits ripened on trees next to the farmhouse, which was surrounded by a shelterbelt. A cow was kept under the same roof and chickens ran about loose in the farmyard, protected by a dog. A beehive hung from the eaves.

All peasants were totally self-supporting and enjoyed the richest and safest possible diet. That they are seen as having been poor and hungry may well reflect the envy of modern man more than anything else. People today have never had the experience of living independently by their own devices, so they know neither spiritual nor material poverty and abundance.

The proof is there before us. Following the war, farms increased steadily in size as the method of cultivation changed, going from 1, to 2, to 4 acres. Although the amount of farmland increased, more and more peasants abandoned farming and left the land. Today, full-time farm households in Japan have grown to 15 and even 25 acres in size, as large as farms in many Western countries. At the same time, they have become increasing unstable and even run the danger of collapse.

Farming operations are usually discussed in terms of economics, but what may appear economically critical is often quite insignificant while something that seems economically trivial may be of overriding importance.

To give one example, the viability of a farming operation is generally deter-mined on the basis of income. Does this make sense? Japan has the world's highest land productivity and output per unit of farmland, but labor productivity and output per farm worker is very low, as is the level of income. Economists have maintained all along that, no matter how high the yields per acre, this means nothing if the remuneration per worker is low. Their ultimate target has been to search for a way of raising income by expanding the scale of operations and raising labor productivity. Granted, Japan's farmers are among the most diligent in the world and with their highly advanced skills and techniques reap high yields. But their small fields make conditions for low-cost farming operations poor. Economically, this means poor productivity of labor and costly farm products that appear to be no match for foreign products.

All of which makes importing and selling foreign farm products, which are relatively inexpensive because of their low production costs, more attractive commercially. The way agricultural scientists and administrators see it, because farming in Japan is economically unjustifiable, we should move toward an international division of labor in food production and have perhaps the United States produce our food for us. This has become the core of Japan's current agricultural policy.

If anything, the low labor productivity of Japan's farmers in spite of their high yields is cause for pride rather than shame. Low income merely indicates either that prices for produce are unreasonably low or that farming equipment and materials are unfairly high, inflating production costs. The farmers have never had any control over the price of farm produce or the costs of materials used in crop production. It is the consumer who determines whether the prices of farm products will be high or low. Farmers never calculated the wages for their labor because farming was done aside from any considerations over money.

Agriculture has nothing to do, fundamentally, with profitability. The overriding concern is how to make use of the land. The goal of farming is to produce a plentiful harvest by bringing out the full forces of nature, because this also happens to be the shortest road to knowing and approaching nature. Farming is not centered on income or on man; at its core are natural fields that transcend man. The fields of nature are the representatives of nature; they are God. The farmer is in the service of God, so immediate gain is a secondary concern. He should rejoice and feel gratified when his fields bear richly.

In this sense, the Japanese farmer, who lived off the smallest possible piece of land, was faithful in bringing out the utmost in both the land and himself. One-acre farmers and quarter-acre farmers are the original image of farming. My proposal for quarter-acre farming is a call to escape from a currency-based economy and devote oneself to fulfilling the true purposes of man.

When I say that crops do not need to be priced, I mean that whether they have prices or not makes no difference to the farmer who dedicates himself to natural farming. Because he has no use for various chemical-based farming materials and does not reckon household labor into his expenses, then his production costs are zero. If all the farmers of the world thought along these lines, crop prices everywhere would settle to the same level and would no longer be needed. Prices are a man-made device; they do not exist in nature. Nature was free, non-discriminating, and fair to begin with. Nothing has less to do with the crops of nature than money.

The price of Japanese rice, the price of Thai rice, and the farmer's price for rice should all be the same. No one should find fault with the shape of a cucumber or the size of a fruit. Bitter cucumbers and sour fruit too each have their proper worth.

What sense is there in importing oranges from the United States and exporting mandarins back? People of each land need only eat food grown close at hand and be contented. What has occurred is that a money-crazed economy has bred senseless competition in food production and thrown dietary habits into chaos.

Agricultural crops grown by natural farming should be assessed on the basis of a natural economy, not a monetary economy. For this to happen, it is necessary that a new system of economics founded on Mu* be developed. Establishing Mu

---

* Nothingness. The origin of all that exists and the antithesis of being.

economics will require us to get rid of our false system of values and unearth the original and true value of agriculture. In addition, Mu natural farming must be supported and implemented with Mu economics and Mu government.

In a nation where everyone tends small farms, circumstances might require that there be some consignment farming, sharecropping on a contractual basis, mutually cooperative cultivation, and even some trading of farm goods grown by natural farming, although this would be limited to the occasional exchange of surplus products on a small scale in open-air markets.

Following the end of the war, Japanese agriculture was regarded as an economic sphere of activity and turned into a business-like occupation. This set a course of destruction from within that has continued unabated ever since. The destruction of an agriculture stripped of its basic meaning has already reached a perilous state.

Economic relief measures are being attempted today, but the most important step that must be taken is not increasing the price of rice. Nor is it lowering the price of materials or cutting back production expenses or raising labor productivity with labor-saving techniques and mechanization or reorganizing the distribution system. None of these are radical measures. Everything depends on whether people are able to return to the viewpoint that "all is unnecessary," that one must "act without acting." Turning around to return to the source of Mu and dedicating ourselves to a Mu economy will not be easy, but it is the only choice we have.

This is the purpose of quarter-acre farming for all the people of the land. If people have a change of heart, they will not need vast green fields to achieve this rebirth; it will be enough for them to work small fields. Our world has fallen into a state of chaos because man, led astray by an embarrassment of knowledge, has engaged in futile labors. The road back to the land, back to the bosom of a pure, innocent nature still remains open to us all.

Epilogue

When the frog in the well gazes outward and observes an image of itself reflected in the mirror of the world, it does not see the mystery of the mirror, but only its distortions and irregularities; it notices only the ugliness and foolishness of its image reflected in the mirror.

Although I could have remained shut up in my own shell and there carried on as I pleased, I thought that I could face the winds of the world and speak freely to all, but I found that I was unable to move.

When I see the flood of fine books in the bookstores, I realize that I too, in my harangues against the worth of books, have been fighting windmills.

Having argued that all is useless ever since I was a youth, I attempted to put my thinking—which denies the understanding of people and posits the principles of "no knowledge," "no worth," and "no action"—into practice through natural farming. My goal was not to compare natural farming, which has no need for human knowledge, with scientific farming, which is.the fruit of human knowledge. The results were already clear for everyone to see.

I was convinced that excellent rice and barley could be grown without doing anything, so it was enough for me to merely grow these. I secretly hoped that, if people saw that I was able to produce rice and barley naturally in this way, then they might reflect upon the meaning of human knowledge and science.

I did not know, however, that the people in today's world are so steeped in scientific and specialized knowledge that they remain unconvinced by such a simple and direct answer. What has surprised me most is that, even when people see the splendid rice and barley that can be grown in a permanently unplowed field without fertilizers or pesticides, even when I explain to them the superiority of natural farming, they hardly seem surprised at all.

People always look at a problem from their own narrow field of specialty or perspective, limiting commentary to an area within which they themselves are capable of analysis and interpretation; they never try to arrive at a conclusion based on comprehensive self-reflection.

Even if it produces splendid rice, most farmers will reject outright a method of rice cropping that leaves even a few weeds standing. Agronomists do not try to spread and popularize herbicides until they are fully effective. What happens to the soil assaulted year after year with these powerful chemicals? The concerns of these people over disease and insect damage also continue to grow without end.

One soil scientist who came to examine my fields was unusual in admonitioning his colleagues that, while it is all right for them to examine the changes in the soil of my fields, they should refrain from criticizing or commenting on the basis of conventional wisdom. He told them that scientists should modestly and quietly observe the changes and that is all. He was one who knows the limits of science.

Most people who see rice and barley grown entirely by the forces of nature feel no sense of wonder. They do not look back on the road I have traveled, and show little interest in the direction in which I am trying to go. All they do is silently examine one instance by the roadside and comment that "this is good" and "that needs improvement."

Yet I am unable to reproach these people. Natural scientists have a great talent for interpreting nature but are hardly ones for approaching and knowing nature. To explain to scientists how natural farming is better than scientific farming would have been an exercise in futility.

People do not have a clear idea of what is natural and what is not. This is why, although they may understand the differences between the shape, form, and methods of natural farming and scientific farming, they are unable to see these as being on entirely different planes and diametrically opposed to each other.

It is a mistake for me to explain the superiority of natural farming to scientists such as these and expect them to reflect on the meaning of science. I might just as well try telling someone from the city who knows nothing of nature that the taste of fresh natural water is better than that of tap water, or telling a sick person that, "Walking is easier than riding a car." To them, whether it is fifty paces or a hundred paces is all the same. This is because they have no idea where the starting point is and are traveling in a different direction.

A true dialogue between man and nature is impossible. Man can stand before nature and talk to it, but nature will not call out to man. Man thinks he can know God and nature, but God and nature neither know man nor tell him anything. Instead, they look the other way.

God and man are travelers passing in opposite directions. Likewise for natural farming and scientific farming. These two paths start from the opposite sides of nature. One seeks to approach closer to nature, the other to move farther away.

Nature on the exterior shows only facts, but says nothing. However these facts are stark and clear. There is no need for explanation. To those who fail to acknowledge these facts, I mutter in my heart, "The farmer is not concerned about high-yield theories and interpretations. What counts is that the yields are the highest possible and the methods used the best there are. This in itself is enough. Certainly you are not telling a farmer to furnish himself the proof with which to convince physicists, chemists, biologists, and specialists in all the other disciplines. And if I had gone to all that ridiculous trouble, this barley you see here would never have been grown. I don't have the time to do research just for research's sake. And to begin with, I don't accept the need to spend one's entire life engaged in such activity."

Nor do I welcome the well-meaning but misguided kindness of some scientists who, wishing to make natural farming universally accepted, try to explain it in scientific terms and support it with theoretical arguments. Natural farming is not a product of the knowledge of clever people. Applying human knowledge and reasoning to natural farming can only distort it, never improve it. Natural farming can criticize scientific farming, but cannot itself be evaluated scientifically.

Back about ten years ago, a large group of specialists, including technical officials from agricultural testing stations in southern Honshu and Shikoku, officials from the Ministry of Agriculture and Forestry, and university scientists from Kyoto and Osaka visited my farm. I told them that, "This field has not been plowed in more than 25 years. Last autumn, I broadcast clover and barley seed over the standing heads of rice. After harvesting the rice, I scattered the rice straw uncut back over the field. I could have sown the rice seed over these heads of barley, except that I sowed the rice seed last autumn together with the barley seed."

Everyone was dumbfounded. As they listened in amazement to me recount how for 25 years I had grown rice and barley in succession by direct seeding and without tilling the soil, how I had relied entirely on grazing ducks to fertilize my field and had never used commercial fertilizers, how I had managed to grow such fine barley without pesticides, some of the assembled scientists grew more and more confused.

But I was delighted to see the reactions of Professor Kawase, an authority on pasture grasses, who was openly impressed with the splendid barley growing among the green manure, and Professor Hiroe, a paleobotanist, who merrily observed a number of different weeds growing at the foot of the barley.

The visitors took photos of the chickens running about the citrus orchard, spun out a *haiku*: "Thickly growing grass/Mandarins luxuriant/So sweet a flavor," and drew sketches of the fresh, green natural orchard. This made my day.

As gorgeous and imposing as are the flowers that bloom on the garden plants that people breed, these have nothing to do with me. Man erred when he tried comparing the flowers created by human intelligence with the weeds. The weeds by the roadside have significance and value as weeds. This is something that cannot be violated or taken away by garden varieties. Let weeds be weeds. Clover belongs to the meadows. Clover has value as clover.

The violet growing along a mountain path blooms for no one in particular, but people cannot overlook or forget it. The moment they see it, they *know*. If people did not change, the world would not change; farming methods would not change.

I am fortunate to have grown rice and barley. Only to him who stands where the barley stands, and listens well, will it speak and tell, for his sake, what man is.

As I look out now at the ripening heads of barley standing golden before me under the sunny May sky, I recall the words of a young visitor from a southern island. After seeing this barley, he left, saying, "I have felt the awesome energy of the earth. What more can I say?"

On the same day, a university professor told me, "It's best to keep philosophy and religion out of the world of science." If the barley had heard, it probably would have answered, "Don't bring science into the world of barley."

Just because science has exploded earlier divinely inspired religious myths, this is no cause for conceit. Science has not overthrown true religion, nor even been able to explain it. What the barley does not tell us is that only religion and philosophy can expose and pass judgment on the horrors of the evils that flow through this world of ours.

In spring, *daikon*, turnip, and rape blossoms bloom beneath the flowering cherry trees. Come the season of the barley harvest, and the sweet fragrance of mandarin flowers drifts over the barley field and out to the Inland Sea. At this time, my natural farm truly becomes a garden of paradise. The young people who come to my farm from the cities live in the crude huts on the mountain among the chickens and goats roaming the orchard. In the evenings, they gather around the sunken hearth and talk and laugh loudly.

I tried to transfer this vision of nature, the fireside chats of these natural people, to the evening conversation of farmers. But my efforts turned out to be nothing more than idle play. Our world of rapid change had no time to lend an ear to the foolish talk of a farmer.

Glossary of Japanese Words

daikon; a large Japanese radish.

ganpi; *Diplomorpha sikokiana*, a thymelaeaceous shrub the bark of which is used to make paper.

hatsutake; *Lactarius hatsudake*, an edible fungus that grows in the shade of pine trees.

hijiki; *Hizikia fusiforme*, an edible brown algae.

hikiokoshi; *Isodon japonicus*, a very bitter perennial of the Mint Family the roots of which are used as a stomachic.

koji; *Aspergillus oryzae*, an amylase-bearing ascomycete mold used in making *miso*.

koshida; *Gleichenia dichotoma*, a fern of the family *Gleicheniaceae*.

matsutake; *Armillaria matsudake*, an edible fungus that grows at the base of the Japanese red pine.

miso; fermented soybean paste.

Mu (無)*;* nothingness or nonbeing; the origin of all that exists and the antithesis of being.

osechi-ryori; Japanese New Year's cooking, consisting of various vegetables and fish boiled down in a sweet sauce.

pachinko; a game similar to pinball, played individually on an upright machine using small metal balls.

sashimi; sliced raw fish.

shiitake; *Cortinellus shiitake*, an edible mushroom widely cultivated in Japan.

shimeji; *Lyophyllum aggregatum*, a highly flavored edible fungus that grows in thick clumps.

tatami; thick straw mats used as flooring in Japanese homes.

urajiro; *Gleichenia glauca*, a fern of the family *Gleicheniaceae*.

wakame; *Undaria pinnatifida*, an edible seaweed of the family *Phaeophyceae*.